UNION INTERNATIONALE DES SCIENCES PRÉHISTORIQUES ET PROTOHISTORIQUES
INTERNATIONAL UNION FOR PREHISTORIC AND PROTOHISTORIC SCIENCES

PROCEEDINGS OF THE XV WORLD CONGRESS (LISBON, 4-9 SEPTEMBER 2006)
ACTES DU XV CONGRÈS MONDIAL (LISBONNE, 4-9 SEPTEMBRE 2006)

Series Editor: Luiz Oosterbeek

VOL. 22

2006 lisboa uispp XV congresso

Humans: Evolution and Environment

C06 - History of Human populations, palaeoecology and ancient DNA
C08 - Bioarchaeology from the Midst of Shells
C14 - Modern Humans origins in Eurasia
C62 - Coastal geoarchaeology: the research of shellmounds
WS32 - Interdisciplinary Studies In Human Evolution

Edited by

Eric Crubezy, Eugénia Cunha and Bertrand Ludes (C06)
Sheila Mendonça de Souza, Eugenia Cunha and Sabine Eggers (C08)
Marcel Otte, Janus Kozlowski and Jean Pierre Bocquet-Appel (C14)
Marisa Coutinho Afonso and Geoff Bailey (C62)
Eugénia Cunha and Group of Studies in Human Evolution (WS32)

BAR International Series 2026
2009

Published in 2016 by
BAR Publishing, Oxford
BAR International Series 2026

Proceedings of the XV World Congress of the International Union for Prehistoric and Protohistoric Sciences / Actes du XV Congrès Mondial de l'Union Internationale des Sciences Préhistoriques et Protohistoriques
Humans: Evolution and Environment

ISBN 978 1 4073 0603 2

Outgoing President: Vítor Oliveira Jorge; Outgoing Secretary General: Jean Bourgeois
Congress Secretary General: Luiz Oosterbeek (Series Editor)
Incoming President: Pedro Ignacio Shmitz; Incoming Secretary General: Luiz Oosterbeek
Volume Editors: Eric Crubezy, Eugénia Cunha; Bertrand Ludes (C06); Sheila Mendonça de Souza, Eugenia Cunha; Sabine Eggers (C08); Marcel Otte; Janus Kozlowski; Jean Pierre Bocquet-Appel (C14) ; Marisa Coutinho Afonso; Geoff Bailey (C62); Eugénia Cunha; Group of Studies in Human Evolution (WS32)

Contacts : Secretary of U.I.S.P.P. – International Union for Prehistoric and Protohistoric Sciences
Instituto Politécnico de Tomar, Av. Dr. Cândido Madureira 13, 2300 TOMAR
Email: uispp@ipt.pt www.uispp.ipt.pt

BAR Publishing is the trading name of British Archaeological Reports (Oxford) Ltd. British Archaeological Reports was first incorporated in 1974 to publish the BAR Series, International and British. In 1992 Hadrian Books Ltd became part of the BAR group. This volume was originally published by Archaeopress in conjunction with British Archaeological Reports (Oxford) Ltd / Hadrian Books Ltd, the Series principal publisher, in 2009. This present volume is published by BAR Publishing, 2016.

Printed in England

BAR
PUBLISHING

BAR titles are available from:

BAR Publishing
122 Banbury Rd, Oxford, OX2 7BP, UK
EMAIL info@barpublishing.com
PHONE +44 (0)1865 310431
FAX +44 (0)1865 316916
www.barpublishing.com

TABLE OF CONTENTS

LIST OF FIGURES

LIST OF TABLES

Session C06

HISTORY OF HUMAN POPULATIONS, PALAEOECOLOGY AND ANCIENT DNA

AUTHENTICITY IN ANCIENT HUMAN MITOCHONDRIAL DNA STUDIES: A REVIEW

Rafael MONTIEL
CIRN and Department of Biology. University of the Azores, Portugal
Rua da Mae de Deus. Apartado 1422, 9501-855 Ponta Delgada, Azores. Portugal
Tel. 00(351)296650476, Fax 00(351)296650100, E-mail: montiel@notes.uac.pt

Abstract: *Research into ancient mitochondrial DNA is plagued by contamination, post mortem damage, and other artefacts. These problems have led to the development of a set of guidelines for authentication of ancient DNA (aDNA) results, including the use of dedicated aDNA laboratories, biochemical preservation tests, multiple negative controls, quantification of the target DNA, cloning and sequencing of polymerase chain reaction products, control of amplicon length, and reproducibility of results. The stringent set of controls suggested by Cooper and Poinar a few years ago is, however, rarely followed in practice, and even when applied carefully, these criteria need not be sufficient to guarantee authenticity. Furthermore, some authors are proposing flexibility and intelligent use of these authenticity criteria, in order to avoid the "checklist" approach, which can result in publication of non-authentic results and in rejection of some interesting and probably authentic ones. On the other hand, although there have been several papers recommending appropriate experimental designs for ancient-DNA studies, there have been few attempts at statistical analysis to establish the confidence of the results obtained. The best way to prevent or minimize contamination is to start precautionary measures as early as possible, ideally starting with sample collection and preparation by field archaeologists, although this is not always possible. Therefore, before starting an aDNA study, a careful experimental design must be devised, a prospective analysis of some samples to determine their usefulness must be performed, and a cost-benefit analysis must be considered.*
Keywords: *Ancient DNA, aDNA, authenticity, Anthropology, mtDNA*

INTRODUCTION

Ancient DNA (aDNA) is defined as DNA recovered from naturally or artificially preserved biological tissues. Two papers published in Nature are considered the starting point of this research field (Higuchi *et al.* 1984; Pääbo, 1985). In a few years, the kind of tissues analysed and their antiquity increased exponentially (Pääbo, 1993). The most spectacular reports presented sequences recovered from samples of million of years old, including insects entombed in amber (DeSalle *et al.* 1992; Cano *et al.* 1993) and dinosaur bones (Woodward *et al.* 1994). Soon after, however, was shown that the recovered sequences were more likely the product of DNA contamination (Walden and Robertson, 1997; Gutiérrez and Marín, 1998, for amber studies; and Allard *et al.* 1995; Hedges and Schweitzer, 1995; Henikoff, 1995; Zischler *et al.* 1995, for dinosaur bones). This issue, already considered a fiasco (Lindahl, 1997), raised several doubts about the entire field. The criticisms were supported by the work of Lindahl (Lindahl and Andersson, 1972; Lindahl and Nyberg, 1972.), who studying the stability of DNA in aqueous solutions concluded that informative segments of DNA couldn't last for more than 100,000 years under any condition (Lindahl, 1993a; 1993b). Although the stability of DNA in time is a polemic issue, now seems clear that if conditions are optimal, DNA may last for 500,000 or even a million years (Hofreiter *et al.* 2001; Poinar *et al.* 1996; 2006). Nevertheless, more than ever it became clear the risk of amplifying exogenous (contaminant) DNA in ancient DNA studies. The severity of this problem may be understood considering that aDNA is strongly damaged and therefore, any trace amount of fresh contaminant DNA may outcompete it in downstream analyses, like molecular cloning or PCR amplification. The problem is exacerbated in ancient human studies, since the researchers belong to the same species as the object of study and the potential sources of contamination increases from step to step of the analysis. Despite the recognized potential of aDNA studies in Anthropology, the ever-present problem of contamination has hampered the application of this technique to the study of ancient human populations. The advice of two of the leading groups in this field is not to work with ancient human samples (Pääbo *et al.* 2004; Willerslev and Cooper, 2005). On the other hand, studies on ancient human populations are still published every year (e.g. Sampietro *et al.* 2005; Alzualde *et al.* 2006; Mooder *et al.* 2006; Töpf *et al.* 2006) and the question remains if the standard authenticity criteria (Cooper and Poinar, 2000) are adequate to ancient human studies.

UBIQUITY OF HUMAN DNA CONTAMINATION

The severity of human DNA contamination was early realized by researchers that amplified human specific mitochondrial DNA (mtDNA) from animal bone extracts (Handt *et al.* 1994; Richards and Sykes, 1995). Recent work has stressed this problem, by detecting human contamination in paleontological teeth samples from animals (Hofreiter *et al.* 2001) and the extensive human contamination that affects animal museum specimens (Malmström *et al.* 2005). The reasoning is that if human DNA is contaminating animal bones, is likely that also may be present in ancient human samples, but in this case, detecting the contaminant sequences is highly problematic if not impossible at all (Gilbert *et al.* 2005b). The problem becomes more serious if the researchers and the samples share the same phylogeographic context (i.e. belong to the same population, or even continent), because in this case researchers and subjects will present similar genetic markers. This shows that ancient human DNA studies require special conditions of analysis and specific authenticity criteria.

THE AUTHENTICITY PROBLEM

The authenticity of a result is defined as the proof that DNA sequences retrieved from ancient samples are in fact from the analysed specimen and not from any other source or artefact. Then, the authenticity problem is mostly related to DNA contamination, but also to post-mortem damage that could produce a sequence different from the actual sequence that the specimen harboured during its life (Gilbert *et al.* 2003; Gilbert *et al.* 2006). As a direct proof of authenticity is hard to establish, the authentication of a result comprises a process that starts with the measures taken during the analysis to avoid the introduction of contaminating DNA. However, most important is to detect contamination that could be introduced during the analysis or, more problematic, the one that could be introduced previously to the arrival of the sample to the laboratory. In order to detect DNA contamination, a series of experimental controls are performed, and replication of results by analysing two or more samples from the same individual are conducted. Nevertheless, in some scenarios, contaminating sequences may not be detected with these controls (Cooper, 1992; 1997) and then, some authenticity criteria must be fulfilled to gain confidence on the results obtained. However, according to some researchers these criteria must be applied in an intelligent way, being more relevant to take a more self-critical approach than just checking criteria off lists (Gilbert *et al.* 2005a).

THE AUTHENTICITY CRITERIA

As soon as the contamination problem was realised, the need of establishing authenticity criteria was recognised. Pääbo *et al.* (1988) proposed that a failure in PCR amplification of big fragments from aDNA extracts could be an additional criterion of authenticity. This early criterion later was developed into the "appropriate molecular behaviour" criterion (see below). However, was in response to Lindahl's criticisms (Lindahl, 1993b) that authenticity criteria gained formal proposals. Handt *et al.* (1994) published the first set of integrated criteria, but were Cooper and Poinar (2000) who published the most comprehensive and stringent list of criteria, further proposing that aDNA studies must fulfil it to be publishable or funded. A third generation of criteria were recently published, relaxing in some way the stringency of the criteria according to the nature and characteristics of the samples (Pääbo *et al.* 2004). Here I review the main criteria and comment about its application on population-level studies.

PHYSICALLY ISOLATED WORK AREA

This is more a procedure to avoid contamination than a criterion of authenticity and consists in separation of the areas according to the analysis performed. DNA extraction must be separated from PCR setup and post-PCR procedures.

NEGATIVE CONTROLS

Either for DNA extraction or amplification, blank controls must be performed. The lower the samples-to-controls ratio, the most reliable will be the procedure. However, it should be considered that clean controls do not imply absence of contamination due to phenomena like the carrier effect (Cooper, 1992), and that contaminated controls also do not imply contaminated samples (e.g. Ivanov *et al.* 1996). Therefore, negative controls are only useful in an integrated authentication framework (Montiel, 2001).

APPROPRIATE MOLECULAR BEHAVIOUR

As aDNA is highly fragmented it is expected to obtain more PCR product for smaller fragments. Also, it is expected to obtain more mtDNA than nuclear DNA, since there are much more copies of mitochondrial than nuclear DNA, increasing the probability of preservation of the former. However, results from some researchers have shown that the interaction of DNA molecules is far more complicated. In some instances it was easier to amplify bigger aDNA than contaminant DNA fragments (Kolman and Tuross, 2000).

REPRODUCIBILITY

At least two extracts of each specimen must be analysed and results must be concordant. In case of disagreement, a third or a fourth sample from the same individual must be characterized. Discordant results may show that contamination has happened, however, concordant results do not warrant authenticity, because if the sample was extensively contaminated, several extracts may contain the same exogenous sequence (Cooper, 1997). This is especially true for samples that do not contain endogenous DNA or have it highly degraded.

CLONING

PCR products (amplicons) must be cloned and multiple clones must be sequenced to assess the integrity of the amplified DNA. This procedure may reveal different sequences amplified at different ratios. Ideally, overlapping fragments must be amplified and cloned. This is one of the must-required proofs of authenticity; however it may render population level studies impractical, especially if the benefit does not justify the costs. According to Bower *et al.* (2005) at least 20 clones must be sequenced to be >95% confident of identifying the most abundant sequence present at 70% in an ancient sample. So, for example, if we must analyse two extracts per individual and amplify (at least) two overlapping fragments, we should sequence at least 80 clones per individual. In a population level study, a minimum of 25 individuals must be analysed and this imply to sequence at least 2000 clones. On the other hand, if the initial template is abundant (>1000 copies), PCR products will be more homogeneous and the cloning procedure does not add new information (Pääbo *et al.* 2004). For example, analysing an exceptionally well-preserved Neanderthal sample, Ovchinnikov *et al.* (2000) obtained the same sequence either by direct sequencing of amplicons or by cloning and sequencing them. Then, the cloning criteria may be relaxed if quantification indicates good preservation, but it is still necessary for exceptional claims (those that may challenge established theories).

INDEPENDENT REPLICATION

This is an extension of the replication inside one lab, but must be performed by a second, (and a third), laboratory. Samples must be sent directly from the museum or collector to avoid transfer of contamination. In population studies, many researchers only replicate a subset of samples (e.g. Sampietro *et al.* 2005; Alzualde *et al.* 2006; Töpf *et al.* 2006).

BIOCHEMICAL PRESERVATION

The samples should present a general good preservation and is advised to test the racemization of amino acids (Poinar *et al.* 1996). Here, the important thing to consider is that an increase in handling of sample implies an increase of contamination risk. In general, if the sample is well preserved, it will contain a good number of endogenous molecules and consequently it will produce a homogeneous result. So in some cases additional handling of samples is not justified, unless the result may be of great relevance. Furthermore, in population-level studies, the percentage of samples that contain amplifiable DNA may be a better indicator of the biochemical preservation.

ASSOCIATED REMAINS

It is advised to analyse fauna to assess human DNA contamination and preservation conditions. However, results in these criteria must be taken with caution; even a good preserved specimen may be contaminated. Moreover, preservation conditions may vary among samples from the same burial. Therefore, unsuccessful attempts to recover DNA from associated remains do not discard preservation in other samples.

RESEARCHER'S CRITERION IN USING THE AUTHENTICITY CRITERIA

In general, the more important the specimen or implications of results, the more stringent the criteria must be to authenticate or justify the results obtained. Samples with good DNA preservation require less stringency, especially concerning cloning and independent reproducibility. In these cases, a subset may indicate that in general the work was conducted correctly. Nevertheless, human population studies require additional criteria, ideally not representing an exponential increase in costs. One of these specific criteria could be to test for mosaic (chimeric) sequences, which represent *in vitro* recombination of different DNA molecules, mainly through "Jumping-PCR" (Pääbo *et al.* 1990). Mosaic sequences may be detected by median networks analysis (e.g. Bandelt, 2005). A mixture of different molecules may be also detected by analysing linked markers. This may be accomplished at the level of the whole mitochondrial genome, by taking advantage of the relationship of markers that define the haplogroups (e.g. Montiel *et al.* 2001).

FUTURE WORK

The criteria of authenticity in aDNA studies must be based in systematic studies assessing the risks of contamination in the different steps of the analysis. In this sense, there are some open questions that must be addressed: i) Probability of amplicon transfer. Many researchers acknowledge that there is a high probability of amplicon transfer, and then laboratory areas are designed to avoid it. One way to control and quantify this probability is to use an internal (modified) control to trace / determine contamination rates. This control may be amplified alongside aDNA samples and then future amplifications of aDNA samples may be routinely tested for the presence of the modified control; ii) Quantification of the amount of contaminant template that may be introduced by handling, taking advantage of identified sequences from archaeologists, for real estimation. To date, only one systematic study addressing this question has been performed, and it shows that of the cloned sequences, 17.13% could be unambiguously identified as contaminants, with those derived from the people involved in the retrieval and washing of the remains present in higher frequencies than those of the anthropologist and genetic researchers (Sampietro *et al.* 2006); iii) Determination of the minimum sub-sample size to be replicated (in a 2nd lab) or further analyzed (i.e. by cloning); iv) Improving of protocols, either for decontamination (e.g. Kemp and Smith, 2005) or searching for novel polymerases insensitive to molecular damage (e.g. Y-family polymerases; McDonald *et al.* 2006); and v) Training of field archaeologists for adequate sample handling for DNA analysis.

Acknowledgments

R.M. is a postdoctoral fellow (SFRH/BPD/32473/2006) of the FCT, Portugal.

References

ALLARD, M.W., YOUNG, D., HUYEN, Y. 1995. Detecting dinosaur DNA. *Science*. **268**: 1192.

ALZUALDE A, IZAGIRRE N, ALONSO S, ALONSO A, ALBARRAN C, AZKARATE A, DE LA RUA C. 2006. Insights into the "isolation" of the Basques: mtDNA lineages from the historical site of Aldaieta (6th-7th centuries AD). *Am J Phys Anthropol*. **130**: 394-404.

BANDELT, H.J. 2005. Mosaics of ancient mitochondrial DNA: positive indicators of nonauthenticity. *Eur J Hum Genet*. **13**: 1106-1112.

BOWER, M.A., SPENCER, M., MATSUMURA, S., NISBET, R.E., HOWE, C.J. 2005. How many clones need to be sequenced from a single forensic or ancient DNA sample in order to determine a reliable consensus sequence? *Nucleic Acids Res*. **33**: 2549-2556.

CANO, R.J., POINAR, H.N., PIENIAZEK, N.J., ACRA, A., POINAR, G.O.Jr. 1993. Amplification and sequencing of DNA from a 120-135 million-year-old weevil. *Nature*. **363**: 536-538.

COOPER, A. 1992. Removal of colourings, inhibitors of PCR, and the carrier effect of PCR contamination from ancient DNA samples. *Ancient DNA Newsletter*. **1**: 31-32.

COOPER, A. 1997. Reply to Stoneking: Ancient DNA – how do you really know when you have it? *Am J Hum Genet*. **60**: 1001-1002.

COOPER, A., POINAR, H.N. 2000. Ancient DNA: do it right or not at all. *Science*. **289**: 1139.

DESALLE, R., GATESY, J., WHEELER, W., GRIMALDI, D. 1992. DNA Sequences from fossil termite in Oligo-Miocene amber and their phylogenetic implications. *Science*. **257**: 1933-1936.

GILBERT, M.T., WILLERSLEV, E., HANSEN, A.J., BARNES, I., RUDBECK, L., LYNNERUP, N., COOPER, A. 2003. Distribution patterns of postmortem damage in human mitochondrial DNA. *Am J Hum Genet*. **72**: 32-47.

GILBERT, M.T., BANDELT, H.J., HOFREITER, M., BARNES, I. 2005a. Assessing ancient DNA studies. *Trends Ecol Evol*. **20**: 541-544.

GILBERT, M.T.P., RUDBECK, L., WILLERSLEV, E., HANSEN, A.J., SMITH, C., PENKMAN K.E.H., PRANGENBERG, K., NIELSEN-MARSH, C.M., JANS, M.E., ARTHUR, P., LYNNERUP, N, TURNER-WALKER, G., BIDDLE, M., KJØLBYE-BIDDLE, B., COLLINS, M.J. 2005b. Biochemical and physical correlates of DNA contamination in archaeological human bones and teeth excavated at Matera, Italy. *J Archaeol Sci*. **32**: 785-793.

GILBERT, M.T., BINLADEN, J., MILLER, W., WIUF, C., WILLERSLEV, E., POINAR, H., CARLSON, J.E., LEEBENS-MACK, J.H., SCHUSTER, S.C. 2006. Recharacterization of ancient DNA miscoding lesions: insights in the era of sequencing-by-synthesis. *Nucleic Acids Research*. doi:10.1093/nar/gkl483.

GUTIÉRREZ, G., MARÍN, A. 1998. The most ancient DNA recovered from an amber-preserved specimen may not be as ancient as it seems. *Mol Biol Evol*. **15**: 926-929.

HANDT, O., HÖSS, M., KRINGS, M., PÄÄBO, S. 1994. Ancient DNA: methodological challenges. *Experientia*. **50**: 524-529.

HEDGES, S.B., SCHWEITZER, M.H. 1995. Detecting dinosaur DNA. *Science*. **268**: 1191.

HENIKOFF, S. 1995. Detecting dinosaur DNA. *Science*. **268**: 1192.

HIGUCHI, R., BOWMAN, B., FREIBERGER, M., RYDER, O.A., WILSON, A.C. 1984. DNA sequences from the quagga, an extinct member of the horse family. *Nature*. **312**: 282-284.

HOFREITER, M., SERRE, D., POINAR, H.N., KUCH, M., PÄÄBO, S. 2001. Ancient DNA. *Nat Rev Genet*. **2**: 353-359.

IVANOV, P.L., WADHAMS, M.J., ROBY, R.K., HOLLAND, M.M., WEEDN, V.W., PARSONS, T.J. 1996. Mitochondrial DNA sequence heteroplasmy in the Grand Duke of Russia Georgij Romanov establishes the authenticity of the remains of Tsar Nicholas II. *Nat Genet*. **12**: 417-420.

KEMP, B.M., SMITH, D.G. 2005. Use of bleach to eliminate contaminating DNA from the surface of bones and teeth. *Forensic Sci Int*. **154**: 53-61.

KOLMAN, C.J., TURROS, N. 2000. Ancient DNA analysis of human populations. *Am J Phys Antrhopol*. **111**: 5-23.

LINDAHL, T. 1993a. Instability and decay of the primary structure of DNA. *Nature*. **362**: 709-715.

LINDAHL, T. 1993b. Recovery of antediluvian DNA. *Nature*. **365**: 700.

LINDAHL, T. 1997. Facts and artifacts of ancient DNA. *Cell*. **90**: 1-3.

LINDAHL, T., ANDERSSON, A. 1972. Rate of chain breakage at apurinic sites in double-stranded deoxyribonucleic acid. *Biochemistry*. **11**: 3618-3623.

LINDAHL, T., NYBERG, B. 1972. Rate of depurination of native deoxyribonucleic acid. *Biochemistry*. **11**: 3610-3618.

MCDONALD, J.P., HALL, A., GASPARUTTO, D., CADET, J., BALLANTYNE, J., WOODGATE, R. 2006 Novel thermostable Y-family polymerases: applications for the PCR amplification of damaged or ancient DNAs. *Nucleic Acids Res*. **34**: 1102-1111

MALMSTRÖM, H., STORÅ, J., DALÉN, L., HOLMLUND, G., GÖTHERSTRÖM, A. 2005. Extensive human DNA contamination in extracts from ancient dog bones and teeth. *Mol Biol Evol*. **22**: 2040-2047.

MONTIEL, R. 2001. Estudio diacrónico de la variabilidad del DNA mitocondrial en población Catalana. Ph.D. dissertation. Autonomous University of Barcelona, Barcelona, Spain. (http://www.tdcat.cesca.es/TDCat-0726101-095837).

MONTIEL, R., MALGOSA, A., FRANCALACCI, P. 2001. Authenticating ancient human mitochondrial DNA. *Hum Biol*. **73**: 689–713.

MOODER, K.P., SCHURR, T.G., BAMFORTH, F.J., BAZALIISKI, V.I., SAVEL'EV, N.A. 2006. Population affinities of Neolithic Siberians: a snapshot from prehistoric Lake Baikal. *Am J Phys Anthropol*. **129**: 349-361.

OVCHINNIKOV, I.V., GOTHERSTROM, A., ROMANOVA, G.P., KHARITONOV, V.M., LIDEN, K., GOODWIN, W. 2000. Molecular analysis of Neanderthal DNA from the northern Caucasus. *Nature*. **404**: 490-493.

POINAR, H.N., HOSS, M., BADA, J.L., PÄÄBO, S. 1996. Amino acid racemization and the preservation of ancient DNA. *Science*. **272**: 864-866.

POINAR, H.N., SCHWARZ, C., QI, J., SHAPIRO, B., MACPHEE, R.D., BUIGUES, B., TIKHONOV, A., HUSON, D.H., TOMSHO, L.P., AUCH, A., RAMPP, M., MILLER, W., SCHUSTER, S.C. 2006. Metagenomics to paleogenomics: large-scale sequencing of mammoth DNA. *Science*. **311**: 392-394.

PÄÄBO, S. 1985. Molecular cloning of ancient Egyptian mummy DNA. *Nature*. **314**: 644-645.

PÄÄBO, S. 1993. Ancient DNA. *Scientific American*. **November**: 60-66.

PÄÄBO, S., GIFFORD, J.A., WILSON, A.C. 1988. Mitochondrial DNA sequences from a 7000-year old brain. *Nucleic Acids Res*. **16**: 9775-9787.

PÄÄBO, S., IRWIN, D.M., WILSON, A.C. 1990. DNA damage promotes jumping between templates during enzymatic amplification. *J. Biol. Chem*. **265**: 4718-4721.

PÄÄBO, S., POINAR, H., SERRE, D., JAENICKE-DESPRES, V., HEBLER, J., ROHLAND, N., KUCH, M., KRAUSE, J., VIGILANT, L., HOFREITER, M. 2004. Genetic analyses from ancient DNA. *Annu Rev Genet*. **38**: 645-679.

RICHARDS, M.B., SYKES, B.C. 1995. Authenticating DNA extracted from ancient skeletal remains. *J Archaeol Sci*. **22**: 291-299.

SAMPIETRO, M.L., CARAMELLI, D., LAO, O., CALAFELL, F., COMAS, D., LARI, M., AGUSTI, B., BERTRANPETIT, J., LALUEZA-FOX, C. 2005. The genetics of the pre-Roman Iberian Peninsula: a mtDNA study of ancient Iberians. *Ann Hum Genet*. **69**: 535-548.

SAMPIETRO, M.L., GILBERT, M.T., LAO, O., CARAMELLI, D., LARI, M., BERTRANPETIT, J., LALUEZA-FOX, C. 2006. Tracking down human contamination in ancient human teeth. *Mol Biol Evol*. **23**: 1801-1807.

TÖPF, A.L., GILBERT, M.T., DUMBACHER, J.P., HOELZEL, A.R. 2006. Tracing the phylogeography of human populations in Britain based on 4th-11th century mtDNA genotypes. *Mol Biol Evol*. **23**: 152-161.

WALDEN, K.K.O., ROBERTSON, H.M. 1997. Ancient DNA from amber fossil bees? *Mol Bio. Evol*. **14**: 1075-1077.

WILLERSLEV, E., COOPER, A. 2005. Ancient DNA. *Proc Biol Sci*. **272**: 3-16.

WOODWARD, S.R., WEYAND, N.J., BUNNELL, M. 1994. DNA sequence from Cretaceous period bone fragments. *Science*. **266**: 1229-1232.

ZISCHLER, H., HÖSS, M., HANDT, O., VON HAESELER, A., VAN DER KUYL, A.C., GOUDSMIT, J., PÄÄBO, S. 1995. Detecting dinosaur DNA. *Science*. **268**: 1192-1193.

IMPROVING INFERENCES ON PAST HUMAN MIGRATIONS: NEW DATA FROM COMPLETE MITOCHONDRIAL SEQUENCING STUDIES

Luísa PEREIRA

IPATIMUP (Instituto de Patologia e Imunologia Molecular da Universidade do Porto), R. Dr. Roberto Frias s/n, 4200-465 Porto, Portugal, lpereira@ipatimup.pt

Abstract: *In the 1990's, uniparental genetic markers contrived the first consistent outlines on the picture of modern human migrations throughout the World. Presently, new technological developments are allowing a much better resolved picture of the human dispersals. For mitochondrial DNA, the highest possible level of molecular resolution is attained, by performing complete sequencing, and a serious effort at nomenclature standardisation produced finely resolved phylogenetic trees. This refined genetic approach allowed already to confirm an extensive human re-settlement of Europe from the Franco-Cantabrian glacial refuge, after the Last Glacial Maximum, improving considerably inferences about European pre-history.*

Key-words: *Complete mtDNA sequence; human population migrations; Last Glacial Maximum; European re-settlement*

Since the first revolutionizing paper, in 1987, by Allan Wilson's group (Cann *et al.* 1987), proving the mitochondrial DNA (mtDNA) potential for inferences on human population history, studies based on this genome have been extensively explored. That first study showed clearly that the most recent common ancestor to all mtDNA lineages, that is "Eve", dated back to around 100.000-200.000 years ago, and that the root of the genetic tree was placed in Africa, being the first solid genetic prove of the Out-of-Africa hypothesis. Now we have a pretty clear idea about the pattern and dates for mtDNA lineage dispersion from Africa to Asia (60.000-70.000 years ago), and hence to Australia (70.000) to Europe (40.000-50.000) and, less consensually, to America (12.000-15.000 or 26.000-34.000). At the European scale, by the end of 2000 (Richards *et al.* 2000), the established picture was that the majority of nowadays mtDNA lineages date back to the Palaeolithic period, with a minor input (around 20%) of Neolithic newcomers, from the Near East, after 10.000 years ago.

By the year 2000, a new era of mtDNA studies initiated. The development of sequencing technologies allowed the surveying of the complete mtDNA molecule (16.569 base pairs) instead of only a small region (around 360 base pairs), so that now all the mtDNA information is ready for analysis. A much-refined description and dating of worldwide human migrations is, consequently, emerging.

In the context of Europe, this strategy has been applied to mtDNA haplogroup V, which represents <5% of European mtDNAs (Torroni *et al.* 2001), and, by us (Pereira *et al.* 2005) and by others (Achilli *et al.* 2004; Loogväli *et al.* 2004), to haplogroup H, by far the most common European haplogroup, which at lower resolution displayed a rather uninformative frequency distribution within Europe (Figure 2.1A). Using polymorphism information derived from the growing complete mtDNA sequence database, we have sequenced 1.580-bp of targeted coding-region segments of the mtDNA genome in 649 individuals harbouring mtDNA haplogroup H from populations throughout Europe, the Caucasus and the Near East. The enhanced genealogical resolution clearly shows that sub-clades of haplogroup H have highly distinctive geographical distributions. The patterns of frequency and diversity suggest that haplogroup H entered Europe from the Near East ~20.000–25.000 years ago, around the time of the Last Glacial Maximum, and some sub-clades (named H1 and H3) re-expanded from an Iberian refugium when the glaciers retreated ~15.000 years ago. Between them H1 and H3 amount to around half of the haplogroup H samples in the database surveyed (Figure 2.1B and 2.1C). They comprise ~65% of haplogroup H lineages in Iberia, ~46% in the north-west, ~27% in central and eastern Europeans, and ~5–15% in the Near East/Caucasus, falling to zero in the Gulf. It is notable that the diversity does not fall within H1 moving from west to east, in accordance with a rapid expansion within the timeframe of the Magdalenian. This shows that a large fraction of the maternal ancestry of modern Europeans traces back to the expansion of hunter–gatherer populations at the end of the last Ice Age.

Besides enhancing the capacity in distinguishing sub-groups inside haplogroups, appearing in different time frames, the complete sequencing also allows to improve date calibration based upon genetic data.

In fact, in the beginning there were some criticisms against the use of HVRI diversity for calculation of expansion times due to its high mutation rate and recurrence of some hypervariable positions (for review see Macaulay *et al.* 1999).

The appearance of the first complete sequences led to the use of (and comparison between) two molecular clocks: one faster based upon the fast-evolving HVRI; and another slower based upon the coding region. Contradicting those fears of the too recurrent HVRI data, when a considerable number of individuals is screened and there is a clear star-like phylogeny for the haplogroup being dated, dates given by both clocks are consistently overlapping (Table 2.1 shows the example for sub-haplogroups belonging to H haplogroup).

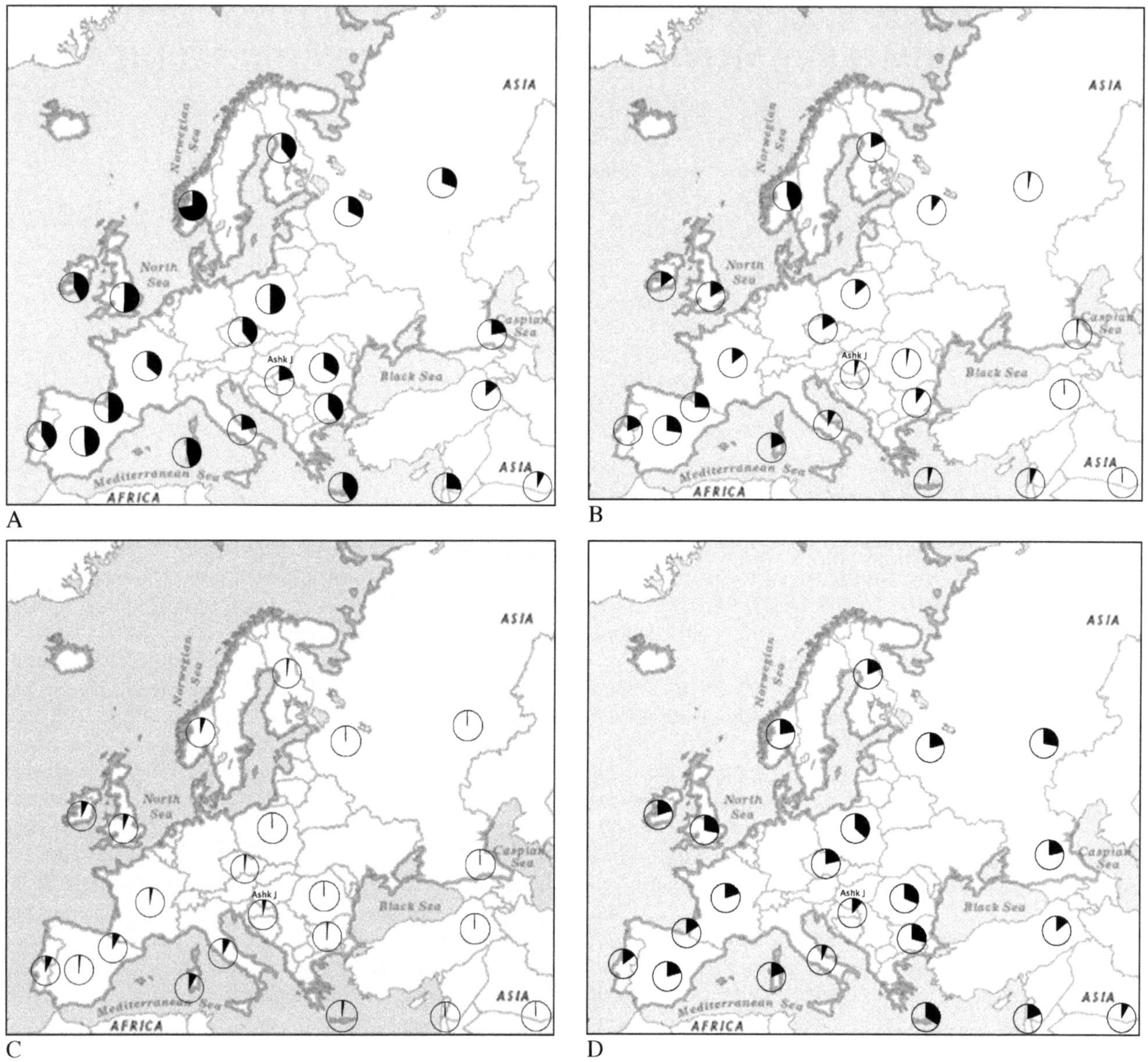

Fig. 2.1. Frequency distributions of haplogroups H (A), H1 (B), H3 (C) and H less H1 and H3 (D) in Europe, the Caucasus and the Near East

Tab. 2.1. Ages for sub-clades of haplogroup H based on coding region (1.580bp) and HVRI (nps 16090–16365) variation. SE means Standard Error

	Age in years (SE)	
Haplogroup (sample size)	Coding region	HVRI
H1 (n = 247)	14.000 (4.000)	16.000 (3.500)
H1 western Europe (n = 195)	14.000 (5.000)	16.000 (3.500)
H1 Iberia (n = 148)	13.000 (6.000)	14.000 (3.000)
H1 eastern Europe (n = 52)	15.000 (6.000)	18.000 (5.000)
H3 (n = 75)	9.000 (3.000)	11.000 (3.000)
All H less H1 and H3 (n = 360)	33.000 (10.000)	23.000 (4.000)

Presently, we are trying to develop further the date calibration by evaluating a complex model of sites partition. Basically, this model allows heterogeneity between different regions in the mtDNA molecule, and a preliminary test (Pereira *et al.* unpublished data) showed that by considering more regions than control versus coding (for instance, by splitting the control in HVRI, HVRII and other; and the coding in 1st codon, 2nd codon, 3rd codon positions, tRNAs and other positions in coding region), standard errors for the expansion times are much more narrowed, and hence, a higher precision in date estimation is attained.

PERSPECTIVES

The near future in mtDNA application to population genetics will consist in a huge effort being put in the characterisation of complete sequence diversity inside specific haplogroups. Focus is being centred in haplogroups implicated in the Out-of-Africa, such as L3, in the origin of all diversity observed outside Africa, and the immediately originated haplogroups M, N and R. The how, the when and the where of the Out-of-Africa can now be addressed much more thoroughly.

We are now in a phase of retrieving all the information possible, based on the extant population diversity, for the maternal lineages dispersal around the World. This was possible due to technological improvements in sequencing techniques. Lets hope a similar technological improvement will allow, in a near future, such a success rate being applied to ancient DNA studies. Then, we will have a clear idea of to what extent present diversity (from the ones reproductively successful) reflects past human diversity (encompassing also extinguished lineages).

Acknowledgments

The sub-H characterisation was possible by a funded Project POCTI/ANT/45139/2002 by Fundação para a Ciência e a Tecnologia (Eixo 2, Medida 2.3 do POCTI, QCA III). IPATIMUP is partially supported by Programa Operacional Ciência, Tecnologia e Inovação (POCTI), Quadro Comunitário de Apoio III.

References

ACHILLI, A. [*et al.*] (2004) The molecular dissection of mtDNA haplogroup H confirms that the Franco-Cantabrian glacial refuge was a major source for the European gene pool. *Am. J. Hum. Genet.* 75: 910–918.

CANN, R.L.; STONEKING, M.; WILSON, A.C. (1987) Mitochondrial DNA and human evolution. *Nature.* 325:31-36.

LOOGVÄLI, E.-L. [*et al.*] (2004) Disuniting uniformity: a pied cladistic canvas of mtDNA haplogroup H in Eurasia. *Mol. Biol. Evol.* 21:2012-2021.

MACAULAY, V.; RICHARDS, M.; SYKES B. (1999) Mitochondrial DNA recombination-no need to panic. *Proc Biol Sci.* 266:2037-2039.

PEREIRA, L. [*et al.*] (2005) High-resolution mtDNA evidence for the late-glacial resettlement of Europe from an Iberian refugium. *Genome Res.* 15:19-24.

RICHARDS, M. [*et al.*] (2000) Tracing European founder lineages in the Near Eastern mtDNA pool. *Am J Hum Genet.* 67:1251-1276.

TORRONI, A. [*et al.*] (2001) A signal, from human mtDNA, of post-glacial recolonization in Europe. *Am. J. Hum. Genet.* 69: 844–852.

Session C08

BIOARCHAEOLOGY FROM THE MIDST OF SHELLS

BIOARCHAEOLOGY OF THE SAMBAQUI GROUPS: SKELETAL MORPHOLOGY, PHYSICAL STRESS AND TRAUMA

Claudia RODRIGUES-CARVALHO
Museu Nacional, Universidade Federal do Rio de Janeiro,
Email: claudia@mn.ufrj.br

Andrea LESSA & Sheila MENDONÇA de SOUZA
Escola Nacional de Saúde Pública Sergio Arouca, FIOCRUZ

Abstract: *The sambaqui people have been studied since the XIX century. A great amount of biocultural information was produced providing a better comprehension of their physical constitution, life style and variability. Sambaqui builders usually had thick limbs, medium to low stature, strong sexual dimorphism (although females were also strong and robust). Studies on acute trauma reveal few fractures and almost no violence. Muscular stress indicators and osteoarthritis data proved a variety of locomotion patterns from one site to other, having in common low asymmetry in limbs' utilization and higher mechanical impact on the upper limbs. Robusticity varies to different groups. The heavier mechanical demands were found in males. Great cemeteries with hundreds of burials, like some sites in Santa Catarina State or small sites concentrations, as seen in some regions of Rio de Janeiro State (among other archaeological data), point to a dense, pacific and well organized society, adapted to the coast, possibly with few specialized activities and low sexual division of labor.*

Resumé: *Les sambaquis sont etudiés dès la fin du siècle XIX. Pendant ce temps furent recueillies des nombreuses informations bio-culturelles que permmetent mieux comprendre ces groupes et sa variabilitée. Les études aux traumas démontrent une basse fréquence de fractures et presque aucune violence. Les signes d'estresse musculaire et l'arthrose signalent une variabilité dans la locomotion. Dans les deux cas, l'asymétrie c'est peu expressive et les membres supérieurs sont plus affectés. La robustesse varie parmis les groupes, mais sa expression dans la majorité des individues est toujour faible où moyenne. L'assemblage de connaissances indique une population bien adaptée aux litoral et a sa exploration.*

INTRODUCTION

The sambaqui people or shellmound builders have been studied since the XIX century, providing an expressive amount of bioarchaeological information (Castro-Faria, 1952; Lima, 1999/2000; Neves, 1988; Prous, 1992). Through this time a set of similarities were observed in many series from sambaquis, such as medium to great robustness, medium to low stature, high sexual dimorphism, pronounced dental wear, low caries rates, and others. These common features leaded to the hypothesis of sambaqui people uniformity what only in recent times hand over to others viewpoints.

In fact, sambaqui builders did share many features in common, as well as similar environments but they also had their own singularities. Those singularities have been better explored by bioarcheologists nowadays. In last decades, new perspectives in bioarcheological studies are detailing biocultural aspects of sambaqui people and reviewing critically the tendency to generalize series-specific data to the whole of sambaqui groups.

Cultural variations along the space and time suggest that these groups did experience distinct lifestyles. This variability possibly implied in singularities in the social organization influencing in daily activities, work division, health status and so forth.

To understand these variations we must improve our knowledge about sambaqui builders' daily life. Several efforts have been made by the last 20 years, in Brazil (as we could see in other presentations of this symposium).

Two important approaches are getting more attention from researches in last decade: the studies about occupational stress related to the effects of regular physical demands and social division of labor; and the studies about interpersonal violence, regarding social tension. The first one evolved from morphological and paleopathological studies. The second is a systematic effort to understand how groups that lived along the coast around 6000 years, deals with conflict intra and extra group.

In this paper we summarize data about sambaqui people physical constitution and describe the state of art of occupational stress and violence studies, trying to synthesize and interpret the similarities and dissimilarities founded in different sambaqui series in South and Southeast Brazil.

PHYSICAL CONSTITUTION OF SAMBAQUI PEOPLE: MORPHOLOGICAL STUDIES

Morphological studies were performed with few interruptions since the XIX century. Craniology (especially craniometry) and questions about origin, migration and admixture of sambaqui groups became the essence of this approach (Alvim & Mello Filho, 1965, 1967/1968; Alvim, Viera & Cheuiche, 1975; Lacerda, 1882, 1885; Neves, 1988; Roquette-Pinto, 1925).

A few studies on post-cranial morphology, performed between the 60's and 80's of the last century, were responsible for the main attempts to reconstruct sambaqui people physical constitution's and were also the main information source about physical demands. These studies revealed great homogeneity in sambaqui populations regarding cranial morphology, stature, sexual dimorphism, robustness and muscle insertion sites development (Alvim & Seyferth, 1968/1969, 1971; Alvim, Vieira & Cheuiche, 1975; Cunha & Alvim, 1971; Uchôa, 1973).

The sambaqui men and women were short, both with thick limbs and marked muscle insertion sites. Although many statures estimations can't be compared, because were made through distinct equations (or using distinct bones), Table 3.1 shows medium values from three sites which date are comparable. These data clearly shows the predominance of low stature among those peoples. Also is clear the stature differences between sexes, with males at least 7 cm taller than females. This difference leads to another characteristic of sambaqui people: the strong sexual dimorphism, observable in general morphology of these groups. Skeleton robustness is also a feature of sambaqui people and although the dimorphism observed in those populations, even women are robust.

Tab. 3.1. Medium values (cm) for stature (femur) in southeast and south sambaqui series

Site/ State	♀	♂	Source
Zé Espinho/RJ	149.0	157.4	Toledo *et al.*, (2006)
Piaçaguera/SP	152.8	160.8	Alvim & Uchôa, 1978
Cabeçuda/SC	149.0	155.7	Alvim & Seyferth, 1971

RJ= Rio de Janeiro State; SP= São Paulo State; SC= Santa Catarina State

This picture of sambaqui people, it's far from being only the result of a common genetic heritage. Certainly, great part of it reflected the cost of a coastal lifestyle, one of the main features shared among those groups.

OCCUPATIONAL STRESS STUDIES

The first systematic studies in articular degeneration (osteoarthrosis) regarding activity interpretations were performed by Brazilian investigators in the 80's, (Machado, 1984; Neves, 1984). Muscle insertion studies just begun in year 2000 (Mata, 2000).

General data about articular degeneration indicates high variability on affected individuals' frequencies, also joint and sex prevalence varies among series (see Tables 3.2 and 3.3 for comparison). The common feature is the scarcity of intense lesions (Machado, 1984, 1992; Neves, 1984; Okumura & Eggers, 2005; Petronilho, 2005; Rodrigues-Carvalho, 2004; Storto, Eggers & Lahr, 1999). The onset of lesions is also variable, but series with other signs of heavy physical activity generally exhibit articular damage in individuals under 30 years old.

Musculoskeletal studies are just beginning in this series, but do show some relevant patterns. Generally, sambaqui people have low occurrence of abnormal ossifications, mainly with mild expressions, according the parameters established by Hawkey & Merbs (1985). Stress lesions also occur in low frequencies but vary in intensity. Mild to moderate robusticity prevailed. The upper limb shows heavier signs of activity, suggesting dominance of water-related activities like swimming, rowing, fishing with nets, etc. There was low occurrence of side dominance what indicates that bilateral activities were probably more frequent or intense than those with side preference.

Tab. 3.2. Occurrence of moderate to severe articular degeneration in apendicular skeleton in four sambaqui series from Rio de Janeiro (at least one joint affected per individual). Source: Rodrigues-Carvalho, 2004. Obs.: N = North; C = Central; S = South

	Moa (N)			Beirada (N)			Zé Espinho (C)			Ilhote (S)		
	N	A	%	N	A	%	N	A	%	N	A	%
♀	12	4	33.3	7	4	57.1	5	2	40.0	8	5	62.5
♂	14	4	28.6	10	2	20.0	8	6	75.0	10	7	70.0

Tab. 3.3. Moderate and severe frequencies of articular degeneration. Sambaqui Jaboticabeira II, Santa Catarina State (south). Source: Petronilho, 2005

Articular group	Moderate		Severe	
	♂	♀	♂	♀
Shoulder	10.00	11.11	30.00	11.11
Elbow	8.33	22.22	16.67	11.11
Wrist	7.69	0.00	0.00	0.00
Hip	0.00	0.00	10.00	10.00
Knee	0.00	0.00	0.00	0.00
Ankle	18.18	25.00	0.00	0.00

Tab. 3.4. Studied sites for acute trauma. Published and unpublished data (papers in preparation)

Site	State	Source
Manitiba	Rio de Janeiro (N)	Machado, 2001
Tarioba	Rio de Janeiro (N)	Machado & Sene, 2001
Beirada	Rio de Janeiro (N)	Machado, 1984; Weska & Lessa (unpublished)
Saquarema	Rio de Janeiro (N)	Kneip, Machado & Crancio, 1995
Zé Espinho	Rio de Janeiro (C)	Lessa *et al.* (unpublished)
Arapuã	Rio de Janeiro (C)	Lessa e Medeiros, 2001
Laranjeiras I	Santa Catarina (C)	Lessa & Scherer (unpublished)
Porto do Rio Vermelho	Santa Catarina (C)	Hubbe, 2005
Jaboticabeira II	Santa Catarina (C)	Storto, Eggers & Lahr, 1999
Forte Marechal Luz	Santa Catarina (S)	Silva & Mendonça de Souza, 1999
Cabeçuda	Santa Catarina (S)	Ferigolo, 1987; Mendonça de Souza, 1995

N = north area; C = central area; S = south area

Regarding sex differences, workload is generally heavier in males than in females, but in some cases women in one site could have activity markers more pronounced that men from other sites (Mata, 2000; Rodrigues-Carvalho, 2004). Marked sex work division was reported only for Sambaqui de Cabeçuda (Santa Catarina State), in this case, low marked deltoideus insertion suggested that rowing was not a regular activity for women (Teixeira, 2004).

As noted in articular degeneration studies, there's a great intra and inter series variability among individuals' robusticity. For example, in a musculoesqueletal stress markers investigation performed with four sambaqui series from Rio de Janeiro State, Rodrigues-Carvalho (2004) founded that people living near calm lagoons, with great amount of resources, had developed less stress markers than people that lived in islands or with intense contact with open sea. In all series studied was also reported considerable variability among individuals from the same sex and age interval, suggesting that activities were not equally distribute to group members.

ACUTE TRAUMA STUDIES

The evidence of acute trauma in prehistoric populations is predominantly represented by accidental fractures, making them especially informative with regard to daily activities and man's interaction with the environment. Although thoroughly published in the international literature, the studies on this kind of lesion among the sambaqui builders are rare.

The first systematic studies about accidental fractures and violence are just beginning. Until a few years ago the data was just mentioned as isolated reports, not investigated as a question itself. Unfortunately, some of these studies did not distributed frequencies of affected individuals by sex, what lead us to aggregate the present data for possible comparisons. Poor skeletal preservation and missing data increase the difficult to compare samples.

A list of sites with accidental and violence-related trauma studies could be seen in Table 3.4. Those preliminary data point out tendencies to be tested in following researches.

Accidental fractures in sambaqui builders

In general, sambaqui builders had low frequency of accidental fractures, indicating a well-succeeded adaptation related to daily activities risk (Table 3.5). This situation could be related to the low land mobility during the subsistence activities due to the great offer of sea and Laguna resources. The higher values in Central region of Santa Catarina State concerns to the exceptional elevated percentual founded in sambaqui Rio Vermelho 2, with 35.7% of individuals with signs of accidental trauma.

Tab. 3.5. Frequency of accidental acute trauma per state/region

State/Region	N	A	%
RJ (North)	44	3	6.8
RJ (Center)	32	5	15.6
SC (North)	10	1	10.0
SC (Center)	26	6	23.0
SC (South)	99	6	6.0
Total	206	21	10.1

According anatomic localization, accidental fracture distribution in sambaqui builders varies. Half individuals were multi-fractured, with no specific pattern of fracture localization. A similar level of injury cicatrisation is not

Tab. 3.6. Accidental fractures distribution among shellmound series

Sites	Upper limb		Lower limb		Multi-fractured	
	A	%	A	%	A	%
RJ (N)	1	5	1	-	1	5
RJ (C)	1	5	2	10	2	10
SC (N)	-	-	1	5	-	-
SC (C)	2	10	-	-	5	25
SC (S)	3	15	-	-	2	10
Total	7	33.3	4	20	10	47.6

compatible with different accidental episodes along the individuals' life, suggesting the fractures occurred during the same event. These features don't suggest accidents during specific activities, which presents specific fracture patterns, but probably during falls, associated to locomotion in irregular areas with rocks and through cliffs and hills, very common in Brazilian shoreline.

Among individuals with just one fracture, upper limbs were more affected especially the clavicles, humeri and radios, once again suggesting the occurrence of falls. The main activities related with regular walks in rough areas would be the collecting shellfish, search for lithic material, use of lithic stations, and control of shoals and environmental changes.

Violence-related trauma in sambaqui builders

The violence-related lesions considered here are the depressed skull fractures, the fractures in the face (especially in the nasal bones) and ulna (parry fracture), and wounds caused by arrow points.

As a rule, the sambaqui groups have low frequency of violence-related trauma. Abundance of resources and a complex ideological and socio-cultural system could have acted as inter-group conflict regulators.

One exception is Zé Espinho sambaqui, with high frequency of lesions (23.8%), which elevates the total frequency for central Rio de Janeiro area groups (15.6%). In this site, the higher frequency was in females (37.5%), suggesting dominance of domestic violence or kidnap. Among males, violence should be associated to inter-group contacts, which is suggested by Tupi-Guarani pottery occurring at the site's upper level.

Violence data comparison: sambaqui builders and other coastal groups

Around 800 A.D. other coastal sites then sambaqui begun to appear. In central area of Santa Catarina State some of these sites are now under investigation (Lessa, 2005; Lessa & Bandeira, 2008, Wesolowski, unpublished), and the preliminary results are discussed here.

Tab. 3.7. Violence-related trauma frequencies in sambaqui builders

	N	A	%
RJ (N)	49	4	8.1
RJ (C)	32	5	15.6
SC (N)	41	2	4.8
SC (C)	26	1	3.8
SC (S)	99	3	3.0
Total	247	15	6.0

These new settlements were shallow, with no intentional shell accumulation and associated with ceramic traditions. In this preliminary approach, lesion comparison between sambaqui builders and groups from the later ceramic settlements point to an increase in violence, especially in males (Table 3.8). This new lifestyle leads to new subsistence strategies and new ideological contexts, which promoted conflict situations. The acquisition of such an important good like pottery must have intensified the changes, robberies and women's kidnapping circuits, contributing to the increasing violence. This is the main hypothesis to be tested in the next investigations (Lessa, 2005).

FINAL REMARKS

Data points out the need of continuous systematic studies considering series distribution over time and space. The hole of investigations summarized here reinforces a model of well-adapted groups to a coastal life-style, with vigorous (but not intensively stressful) activity pattern. This model includes a low pattern of accidental trauma.

Sambaqui groups' social structure would be efficient enough to buffer interpersonal conflicts. Great cemeteries with hundreds of burials and huge mounds, like some site in Santa Catarina State or concentrations of small sites, as seen in some regions of Rio de Janeiro State, among other archaeological data, point to a dense pacific and well organized society. These features changed in later

Tab. 3.8. Violence-related trauma frequencies in later settlements groups

	Male			Female			Total		
	N	n	%	N	n	%	N	n	%
Tapera	29	5	17.2	26	1	3.8	55	5	10.9
Itacoara	11	3	27.2	8	2	25	19	5	26.3
Enseada (upper level)	5	1	20	4	-	-	9	1	11.1
Total	45	9	20	38	3	7.8	83	11	13.2

sambaqui occupational period, a possible result of interaction with ceramist groups that were moving towards the coast. It's not clear yet that violence was the major force on the replacement of sambaqui people by others groups, but its presence is now unquestionable in this moment in our prehistory.

References

ALVIM, M.C.M. & MELLO FILHO, D.P., 1965. Morfologia Craniana da população do Sambaqui de Cabeçuda (Laguna, Santa Catarina) e sua relação com outras populações de paleoameríndios do Brasil. In: sn, *Homenaje a Juan Comas em su 65 Aniversario*, Mexico, vol. 2, pp 359-366.

ALVIM, M.C.M. & MELLO FILHO, D.P., 1967/68. Morfologia da população do Sambaqui do forte Marechal Luz (Santa Catarina). *Revista de Antropologia*, 15/16: 1-12.

ALVIM, M.C.M. & SEYFERTH, G., 1968/1969. Estudo morfológico do úmero na população do Sambaqui de Cabeçuda (Laguna, Sta. Catarina). *Revista do Museu Paulista*, nova série, XVIII:119-126.

ALVIM, M.C.M. & SEYFERTH, G., 1971. O fêmur na população do Sambaqui de Cabeçuda (Laguna, Estado de Santa Catarina, Brasil). Estudo morfológico e comparativo. *Boletim do Museu Nacional*, nova série, 24:1-14.

ALVIM, M.C.M.; VIEIRA, M.I. & CHEUICHE, L.M.T., 1975. Os construtores dos sambaquis de Cabeçudas, SC e de Piaçaguera, SP: estudo morfológico comparativo. *Arquivos de Anatomia e Antropologia*, 01:393-406.

CASTRO-FARIA, L. *Pesquisas de Antropologia Física no Brasil*. In: Boletim do Museu Nacional, nova série, antropologia, Rio de Janeiro, 13:1-106, 1952.

CUNHA, E.S. & ALVIM, M.M.C., 1971. Contribuição para o conhecimento da morfologia das populações indígenas da Guanabara. Notas sobre a população do sítio arqueológico Cabeça de Índio. In: Instituto de Pré-História (Org.), *O Homem antigo na América*, pp 21-24.

FERIGOLO, J.; 1987 *Paleopatologia comparada de vertebrados; "Homem de Lagoa Santa", "Homem do Sambaqui de Cabeçuda" e mamíferos pleistocênicos*. Tese de Doutorado. Porto Alegre: Departamento de Paleontologia e Estratigrafia; Universidade Federal do Rio Grande do Sul.

HAWKEY, D.E. & MERBS, C.F., 1995. Activity-induced musculoskeletal stress markers (MSM) and subsistence strategy among ancient Hudson Bay Eskimos. *International Journal of Osteoarchaeology*, 5: 324-338.

HUBBE, M. 2005. *Análise biocultural dos remanescentes ósseos humanos do Sambaqui Porto do Rio Vermelho 02*. Doctoral Thesis. Instituto de Biociências da Universidade de São Paulo.

KNEIP, L.M.; MACHADO, L.C.; CRANCIO, F. 1995. Ossos humanos trabalhados e biologia esqueletal do Sambaqui de Saquarema, RJ. In: L.M. Kneip (Ed.) *Documento de Trabalho* nº 3, Série arqueologia Rio de Janeiro: UFRJ.

LACERDA, J.B., 1882. A morfologia craneana do homem dos sambaquis. *Revista da Exposição antropológica Brasiliera*, 4:133 -137.

LACERDA, J.B., 1882. O Homem dos sambaquis. *Archivos do Museu Nacional*, 6:175 -203.

LESSA, A. 2005. Reflexões preliminares sobre paleoepidemiologia da violência em grupos ceramistas litorâneos: (I) Sítio Praia da Tapera – SC. *Revista do MAE* 15:199-207.

LESSA, A.; MEDEIROS. J. 2001. Reflexões preliminares sobre a questão da violência em populações construtoras de sambaquis: análise dos sítios Arapuá (RJ) e Cabeçuda (SC). *Revista do MAE* 11: 77-93.

LIMA, T.A., 1999/2000. Em busca dos frutos do mar: os pescadores-coletores do litoral centro-sul do Brasil. *Revista USP*: 270-327.

MACHADO, L.C., 1984. *Análise de Remanescentes Ósseos Humanos do Sítio Arqueológico Corondó, RJ. Aspectos Biológicos e Culturais*. Série monografias. Rio de janeiro: Instituto de Arqueologia Brasileira.

MACHADO, L.C. 2001. Estudo de condições patológicas na população esqueletal do Sambaqui da Beirada, Saquarema, Estado do Rio de Janeiro. *Resumos do III Congreso Latinoamericano de Antropologia Biológica*, Rio de Janeiro, p. 28.

MACHADO, L.C.; SENE, G.M. 2001. Simbolismo, identidade cultural e análise biológica: os remanescentes esqueletais humanos do Sambaqui da Tarioba.

In O. Dias JR; J. Decco; M. Froés (orgs.) *A pré-história de Rio das Ostras: sítio arqueológico Sambaqui da Tarioba*. p. 51-79.

MENDONÇA DE SOUZA, S.M.F. 1995. *Estresse, Doença e Adaptabilidade. Estudo Comparativo de Dois Grupos Pré-históricos em Perspectiva Biocultural*. Tese de Doutorado. Rio de Janeiro: Escola Nacional de Saúde Publica/FIOCRUZ.

MATA, A.L., 2000. *Marcadores de Stress Músculo Esqueletal em uma Población Prehistórica Brasileira*. Specialization Monograph (C. Rodrigues-Carvalho, orientadora), Specialization course in Paleopathology, History and Human Disease Evolution/Curso de Especialização em Paleopatologia, História e Evolução das Doenças Humanas. Rio de Janeiro: Escola Nacional de Saúde Pública/Fiocruz.

NEVES, W.A., 1984. Incidência e distribuição de osteoartrites em grupos coletores de moluscos do litoral do Paraná: uma hipótese osteobiográfica. *Clio, série Arqueológica*, 1(6):47-62.

NEVES, W.A. *Paleogenética dos grupos pré-históricos do Litoral Sul do Brasil (Paraná e Santa Catarina)*. Pesquisas, Antropologia, nº 43, 1988.

OKUMURA, M.M.M. & EGGERS, S., 2005. The people of Jabuticabeira II: reconstruction of the way of life in a Brazilian shellmound. HOMO, 55: 263-281.

PETRONILHO, C. 2005. Comprometimento articular como um marcador de atividade em um grande sambaqui cemitério. Dissertação de Mestrado. Universidade de São Paulo, São Paulo.

PROUS, A. 1992. Arqueologia Brasileira. Brasília: Ed. UnB.

RODRIGUES-CARVALHO, C. 2004. Marcadores de Estresse Ocupacional em Populações Sambaquieiras do Litoral Fluminense. Tese de Doutorado.

ROQUETTE-PINTO, E. 1925. Nota sobre o material anthropológico do Sambaqui de Guaratiba. *Boletim do Museu Nacional*, 1: 397-399.

SILVA, E.; MENDONÇA DE SOUZA, S. 1999. Paleopatologia da população pré-histórica do Sambaqui de Forte Marechal Luz: reabertura de arquivo. *Anais da X Reunião da Sociedade de Arqueologia Brasileira*, Recife. CD-Room.

STORTO, C.; EGGERS, S.; LAHR, M. 1999. Estudo preliminar das paleopatologias da população do Sambaqui Jaboticabeira II, Jaguaruna, SC. *Revista do Museu de Arqueologia e Etnologia da Usp* 9: 61-71.

TEIXEIRA, A. 2004. Estudo de Marcadores de Estresse Musculoesquelético em Restos Esqueléticos dos Habitantes do Sambaqui de Cabeçuda, Laguna, SC. – Uma Abordagem Biocultural. Disertação de Mestrado. Instituto de Ciências Biomédicas/UFRJ, Rio de Janeiro.

UCHÔA, D.P. 1973. Arqueologia de Piaçaguera e Tenório: análise de Dois sítios Pré-cerâmicos do Litoral Paulista. Tese de Doutorado. Faculdade de Filosofia, Ciências e Letras de Rio Claro. 230pp.

WESOLOWSKI, V.; LESSA, A. & BANDEIRA, D. 2008. Marcas da violência: Estudo de caso de um esqueleto "flechado" no litoral norte de SC. Resumos do V Workshop Arqueológico de Xingó, Aracaju.

COASTAL VERSUS RIVERINE SHELLMOUND BUILDERS IN BRAZIL: METHODOLOGICAL ISSUES REGARDING BIODISTANCE

Maria Mercedes M. OKUMURA*#, Ligia G. BARTOLOMUCCI*, José FILIPPINI*, Rita VARGIU+, Sabine EGGERS*

* Instituto de Biociências, Universidade de São Paulo, Brazil

Leverhulme Centre for Human Evolutionary Studies, University of Cambridge, UK

+ Universitá di Roma "La Sapienza", Italy

Abstract: *Studies regarding biodistance between Brazilian coastal and riverine shellmound inhabitants have pointed to distinct conclusions about this relation. Craniometrical analyses and analyses based on non-metrical dental traits point to some similarities between coastal and inland (riverine) individuals. However, non-metrical cranial traits indicate that there is some differentiation between coastal and riverine individuals. This paper aims to discuss how different methodological approaches on biodistance can provide different answers to the same problem.*

Key Words: *Ribeira de Iguape valley, craniometry, non-metrical traits, shellmound, Brazil*

Resumo: *Estudos de biodistância entre indivíduos oriundos de sambaquis costeiros e fluviais (interioranos) têm resultado em conclusões distintas. Análises craniométricas e análises de caracteres não métricos dentários entre esses dois grupos mostram uma certa similaridade entre os indivíduos costeiros e fluviais. No entanto, análises baseadas em caracteres não métricos cranianos indicam que há uma diferenciação substancial entre os grupos da costa e do interior. Este trabalho pretende discutir como diferentes abordagens metodológicas em biodistância podem apontar para respostas distintas de uma mesma pergunta.*

INTRODUCTION

Coastal shellmounds or "sambaquis" are archaeological sites, sometimes reaching 30 m height scattered along almost the entire shoreline of Brazil. They consist of layers of shells intermingled with faunal remains, artefacts, burials and other remains (Edwards *et al*, 2001; Gaspar, 1998). Riverine shellmounds are located inland, are much smaller and consist basically of the same types of remains. For a more detailed comparison in terms of implantation, temporal and geographical distribution of these coastal and riverine shellmounds refer to the paper in this issue called "Riverine versus coastal shellmounds in Brazil" by Eggers. Some riverine shellmounds are older than the oldest coastal sites registered until now in Brazil, but all are dated from the middle Holocene onwards.Biodistance studies are based on the measurement of the biological relation between groups of people based on the quantitative analysis of skeletal traits (Buikstra *et al*, 1990). Groups sharing similarities are more closely related than populations with many differences.Basically, the ideas concerning the reasons for morphological features to vary among humans are two: either they are genetically determined and driven by natural selection or they are a result of non-genetic interactions between bones and environment. Obviously, none of these visions is entirely correct. At the present, cranial features like size and shape are considered to be mainly genetically controlled, but formerly they were thought to be subject to major environmental modifications (Hiernaux, 1963). One of the most important studies about the environmental influence on cranial features was performed by Franz Boas (1966) on European immigrants in United States. Boas concluded that cranial shape was very sensitive to environmental changes; therefore, cranial plasticity should not be ignored in craniometrical studies. This result has been cited for more than 90 years, being the major argument against the use of cranial measurements as a way to unravel biological affinities. However, the recent reassessment carried out by Sparks & Jantz (2002) of the 8.000 skulls studied by Boas, but within a modern statistical and quantitative genetic framework, no longer supports the major role of environment in cranial size and shape. The genetic evidence of the origin of the non-metrical cranial traits is indirect, and based on experimental data on animal models, association of some traits with hereditary syndromes, and distinct frequencies of these characters in different human groups (Hauser & De Stefano, 1989). Furthermore, non-metrical trait distribution among humans agrees with the distribution found for Mendelian traits like blood groups, and distance matrices based on craniometrical analysis. Another line of evidence is provided by the study of non-metrical dental traits, which may contribute to the reconstruction of biological relationships among groups. Dental traits appear to have considerable genetic components in their determinations (Scott, 1973; Nichol, 1989; Scott & Turner II, 1997) which makes them amenable to micro-evolutionary analyses (Irish 2000; Guattelli-Steinberg *et al*, 2001; Irish & Guattelli-Steinberg, 2003). Furthermore, teeth are not subjected to remodeling caused by aging processes, except for masticatory surface abrasion. This makes it possible to score dental traits both in infant and adult individuals, so that larger population samples can be analyzed so that larger population samples can be analysed.The objective of this study is to discuss what can be learned through the use of different biodistance approaches (metrical and non-metrical traits) applied to the study of human skeletal remains from coastal and riverine

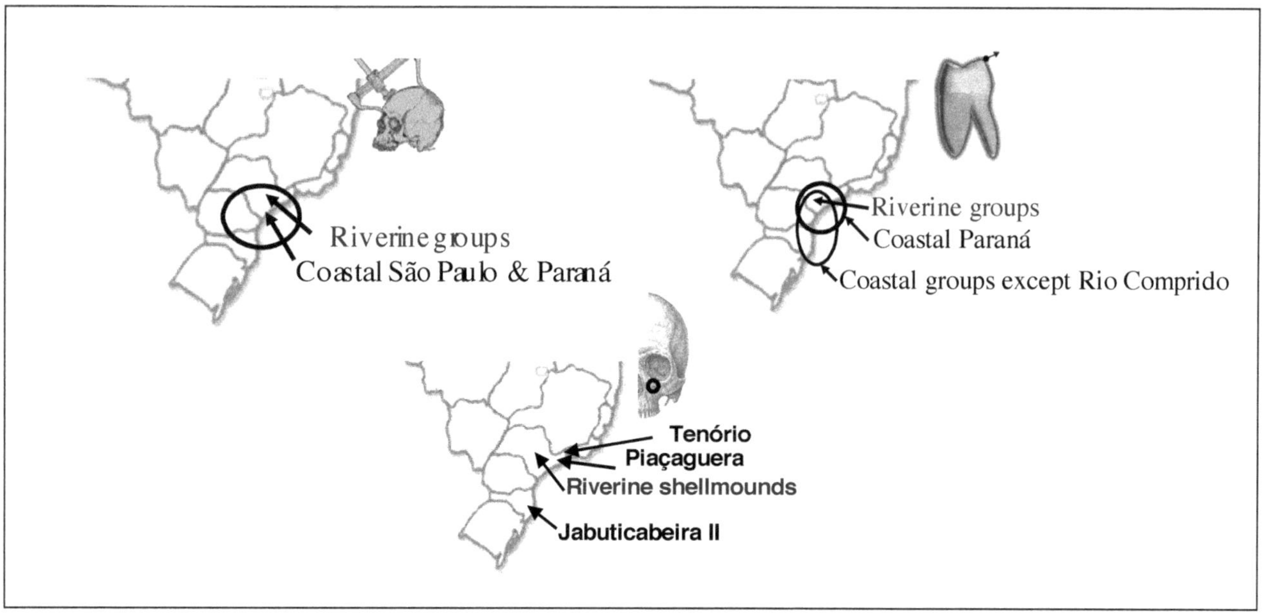

Fig. 4.1. Summary of results using different biodistance approaches regarding the comparison between riverine and coastal shellmounds. Craniometrical and some non-metrical dental analysis resulted in similarities (circles) between the riverine shellmounds in São Paulo and the neighbouring coastal sites in Paraná and São Paulo. Results based on non-metric cranial traits show significant differences between riverine shellmound dwellers and coastal sites in São Paulo (Tenório and Piaçaguera sites) and Santa Catarina (Jabuticabeira II)

shellmounds in Ribeira de Iguape valley (in the boundary between São Paulo and Paraná States), Brazil.

MATERIAL & METHODS

We used metrical cranial traits, and non-metrical traits from the skull and the teeth to estimate the interpopulational variation spatially as a parameter of genetic similarity or divergence (Hauser & De Stefano, 1989).

Metrical traits are continuous variables obtained from linear measurements (length, width, projections or radius) that are used to characterize the size and shape of skeletal elements, most of the times, the skull (Larsen, 1997). The quantitative study of size and shape is always implicit in any craniometrical approach, and in order to try to analyse them separately, it is necessary to apply a correction (Darroch & Mosimann, 1985). Several landmarks are used to delimit the distances measured for craniometrical studies. The landmarks are points located on the cranium that present some anatomical, functional, structural developmental or evolutionary meaning.

Non-metrical cranial studies are based on the observation of the presence or absence of anatomical characters. There are basically four types of non-metrical traits: a) ossicles and supranumerary bones, such as the lambdoid ossicles; b) hyperostosic traits, such as the palatine torus; c) hypostosic traits like the Huschke's foramen and d) foramen and suture variation. These traits vary in number and / or localization.

Non-metrical dental traits behave basically as do the non-metrical cranial traits, in that they are determined genetically. Morphological dental traits have proved to be highly discriminating in the phenetic analysis among populations because of their abundance, strong hereditary component and low environmental influence. They refer to number of cusps and roots, and shape of the tooth crown recorded according to the Arizona State University Dental Anthropology System (ASUDAS) using rank-scale reference plaques (Turner II *et al.*, 1991).

Each of the different biodistance approaches will be assessed separately. Craniometrical data will be summarized here, since there is a detailed article in this issue about it (Okumura & Neves: "The contribution of cranial morphology of human skeletal remains to the understanding of the biological affinities between coastal and riverine shellmounds in Southern Brazil"). For the craniometrical study 9 males and 3 females from 5 riverine shellmounds from Ribeira de Iguape valley (Moraes, Itaoca 1, Estreito 1, Pavão III and Pavão XVI), dated to between 5.900 and 1.200 yBP were analyzed. They were compared to prehistoric groups (131 males and 94 females) from inland and coastal neighbouring regions using Principal Component Analysis (PCA) based only on cranial shape.

For the non-metrical cranial studies 3 males, 4 females and 1 adult of unknown sex from the riverine shellmound Moraes (that yielded the greatest number of individuals), dated to between 5.900 and 4.600 yBP were analyzed. They were compared to 22 males, 16 females and 6 adults of unknown sex from 3 coastal shellmounds from São Paulo (Piaçaguera, Tenório) and Santa Catarina State

Tab. 4.1a. Mean Measure of Divergence (MMD) for non-metrical cranial traits. Riverine Moraes is different from coastal sites from São Paulo (SP) and Santa Catarina (SC).
Legend: ns: non significant; * significant differences

	All coastal sites	Piaçaguera (SP)	Tenório (SP)	Jabuticabeira II (SC)
Moraes (SP)	*	*	*	*
Piaçaguera (SP)			ns	*
Tenório (SP)				ns

Tab. 4.1b. Mean Measure of Divergence (MMD) for non-metrical dental traits with sites grouped together. Riverine shellmound dwellers are different from coastal groups from Santa Catarina, but are similar to those in Paraná. Legend: ns: non significant; * significant differences

	Coast of Paraná	Northern coast of Santa Catarina	Southern coast of Santa Catarina
Riverine composite	ns	*	*
Paraná		*	*
North S Catarina			*

(Jabuticabeira II), dated to between 5.000 and 1.900 yBP. One hundred and twenty five variables were initially collected. Statistical procedures included the estimation of the Mean Measure of Divergence (MMD) (Smith, 1977) and Sanghvi's distance. After eliminating correlating traits and verifying that sex was not significantly associated with any of the remaining traits, the homogeneity of the 4 samples was tested using the Jaccard index. As they were homogeneous to more than 75% the actual statistical analysis could be carried out. The final number of variables used for non-metrical cranial biodistance studies was 27. Details of these procedures can be found in Filippini & Egger, 2005/2006.

For the non-metrical dental studies 22 adults from 3 riverine shellmounds from Ribeira de Iguape valley (Moraes, Pavão XVI, Estreito), dated to between 5.900 and 1.600 yBP were analyzed. They were compared to 215 individuals from coastal shellmounds of the regions Paraná (Guaraguaçu, Matinhos), Northern Santa Catarina (Morro do Ouro, Enseada, Rio Comprido) and Southern Santa Catarina (Jabuticabeira II), dated to between 4.900 and 1.500 yBP, using Mean Measure of Divergence (MMD) and Sanghvi's distance. The MMD coefficient was considered significant at the 0.05 probability level when it was greater than twice its standard deviation (Sjøvold, 1977). Although Itacoara is a riverine site, it could not be placed within the group of riverine shelmounds from Ribeira de Iguape valley because it is located in Santa Catarina.

RESULTS

The craniometrical study shows that the male riverine shellmound dwellers from Ribeira de Iguape valley are more closely associated to the coastal shellmound males from the Southern and Central seashore from São Paulo (Figure 4.1). Among females we observe that the women from the riverine shellmound Moraes presents more affinity with the neighbouring women from Paraná and São Paulo. Therefore, PCA pointed to a certain degree of similarities between the riverine groups from Ribeira de Iguape valley and the coastal groups from São Paulo State and Paraná (the neighbouring state).

The MMD applied to non-metrical cranial data showed that riverine shellmound people are significantly different from those who inhabited the coast of São Paulo and Santa Catarina (Table 4.1a). Some differences also were observed among the coastal shellmounds Piaçaguera (São Paulo) and Jabuticabeira II (Santa Catarina). These results were confirmed by the Sanghvi's distance regarding the affinities between the riverine shellmound and the coastal ones. However, Sanghvi's distances within coastal shellmounds were different from those obtained from MMD. Taking both these non-metrical cranial procedures into account (MMD and Sanghvi's distance), riverine shellmound builders from Moraes are morphologically different in comparison to coastal shellmound builders from São Paulo and Santa Catarina.

For the sake of comparison with the craniometrical data, the sites studied using dental non-metrical data were divided into the same geographical regions (Table 4.1b). According to the results of the MMD of non-metrical dental data, riverine shellmound builders are morphologically similar to the coastal shellmound dwellers from Paraná, but different from those who built the sites in Santa Catarina. However, when analysing the Sanghvi's distance results, riverine dwellers do not appear more similar to the coastal neighbours from Paraná than to the other coastal sites in Santa Catarina.

When analysing the sites separately to ensure better comparability with the non-metrical cranial study, taking

Tab. 4.1c. Mean Measure of Divergence (MMD) for non-metrical dental traits applied to individual sites. Riverine Moraes is significantly different from coastal Rio Comprido, but is similar to the other sites from São Paulo (SP), Paraná (PR) and Santa Catarina (SC). Legend: ns: non significant; * significant differences

	Rio Comprido (SC)	Moraes (SP)	Itacoara (SC)	Matinhos (PR)	Morro do Ouro (SC)	Enseada (SC)
Moraes (SP)	*					
Itacoara (SC)	ns	ns				
Matinhos (PR)	*	ns	ns			
Morro do Ouro (SC)	ns	ns	ns	ns		
Enseada (SC)	*	ns	ns	ns	*	
Jabuticabeira II (SC)	ns	ns	ns	ns	*	*

into consideration only those with a better sample size, another picture emerges. Using MMD the riverine dwellers from Moraes are only significantly different from the coastal people from Rio Comprido (Santa Catarina), but are similar to the other coastal sites, including Matinhos in coastal Paraná (Table 4.1c). When Sanghvi's distance is applied to these sites, a different picture emerges. Moraes seems to be very near to or even inside the dispersion of coastal sites. Jabuticabeira II (and not Rio Comprido) is now the most distant site from the riverine shellmound Moraes.

DISCUSSION

Craniometrical studies show that riverine shellmound dwellers are similar to their neighbour coastal groups in São Paulo, but non-metrical cranial data disagree with that. However, when analysing non-metrical dental traits, the riverine groups are similar to coastal groups from the neighbour state Paraná, and in some analyses also to coastal sites in general (except Rio Comprido). Thus, there are different results according to distinct biodistance approaches.

However, each study used a different set of coastal sites for comparison with the riverine dwellers from the Ribeira de Iguape valley. One can imagine that if the dental study had included coastal sites from the state of São Paulo, the riverine shellmound groups would possibly be also similar to them, as they are when considering craniometrical data. It also is clear that when only a few sites are studied differences emerge between the sites, which tend to disappear when more sites are considered. The reason why the different approaches presented here used different sets of sites was that the studies were undertaken individually and access to the collections was different for each of them. Additionally, the studies with a smaller number of sites are Master thesis (Filippini, 2004; Filippini & Eggers, 2005/2006; Bartolomucci, 2006) while the one with the greatest number of sites is a PhD dissertation (Okumura, 2008). What should and will be done in the future is to reassess the collections of all coastal sites and collect the non-metrical cranial as well as non-metrical dental data that is lacking. Probably then, the results do get more concordant.

On the other hand, as seen in some reports using concomitantly mitochondrial and nuclear DNA, as well as Y-chromosome analyses (Garrigan & Hammer, 2006), even when using exactly the same set of individuals, grouping them in the same manner and having a very good sample size, the results concerning different markers may not be entirely coincident. This does not mean that one approach is better than the other, but it reflects the different evolutionary trajectories and selective forces acting upon each marker and the human groups.

Climate, microenvironment, resources and population size in coastal areas must have been different than those found at the riverine sites at Ribeira de Iguape valley. The way human groups responded and adapted to these different selective pressures may be the cause of divergent results according to the methods used in biodistance studies. However, the contact between the riverine shellmound dwellers from the Ribeira de Iguape valley and coastal groups from neighbouring regions must have caused the reduction of these differences, as we have seen through the craniometrical and dental data. Indeed, the contact between the coast and the inland dwellers could have been facilitated in the Ribeira de Iguape valley because of its geomorphological features.

There is still another reason that could have contributed to the different results of these biodistance studies. Different selective forces as well as diverse evolutionary trajectories could not only have acted upon the populations themselves, but also on the different markers used for biodistance estimative (Hauser & De Stefano, 1989). It is obvious that all these different markers cannot be completely selectively neutral. In the contrary, we know that many of them must have been under strenuous selective forces. Examples of that relate to cranial capacity, degree of prognatism, relative size of posterior teeth in relation to anterior teeth and as such were the result of alterations in cranial and dental measurements. Other cranial and dental traits may, in the contrary, have been subjected to only minor selective forces.

Finally, there are also important differences in the approaches used that have to be kept in mind when considering the availability of the material. Craniometrical analyses needs well preserved and complete crania, on the other hand, continuous variables allow more statistical procedures. Non-metrical traits have the advantage over craniometrical ones that incomplete crania can be used. However a greater sample size is necessary for proper statistical analysis. Dental studies also bear the advantage that incomplete human remains can be studied and especially those with loose teeth. However, when the collection of interest has most of the teeth in the alveoli, this eliminates the possibility of studying the variation in root morphology.

Thus, each method has its advantages and its drawbacks and according to the question asked and the collection available one or the other may be more efficient in the first moment. However, the more types of variables and approaches we use in a careful and thoughtful way, the more information we get on the processes involved in the history of our species.

Acknowlegdements

Thanks are due to the following institutions and their corresponding staff: Museu Nacional (UFRJ-Rio de Janeiro), Museu de Arqueologia e Etnologia (USP-São Paulo), Museu Paranaense (UFPR-Paraná), Centro de Estudos e Pesquisas Arqueológicas (UFPR-Paraná), Museu de Arqueologia e Etnologia (UFPR-Paraná), Museu Arqueológico de Sambaqui de Joinville (MASJ-Santa Catarina), Museu Arqueológico do Rio Grande do Sul (MARSUL-Rio Grande do Sul), Zoological Museum (University of Kopenhagen). Financial support was received from FAPESP (Grants 02/13441-0 and 04/01321-6), FAPESP-CEPID, CNPq (305918/00-85).

References

BARTOLOMUCCI, L.B.G. (2006) – *Variabilidade biológica de sambaquieiros através de morfologia dentária*. MSc thesis, Universidade de São Paulo, Brazil.

BOAS, F. (1966) – *Race, Language and Culture*. New York: The Free Press.

BUIKSTRA, J.E.; FRANKENBERG, S.R.; KONIGSBERG, L.W. (1990) – Skeletal biological distance studies in American physical anthropology: recent trends. Am J Phys Anthrop. 82:1, p. 1-7.

DARROCH, J.N.; MOSIMANN, J.E. (1985) – Canonical and principal components of shape. *Biometrika*. 72: 2, p. 241-252.

EDWARDS, H.G.M. [*et al.*] (2001) – Raman spectroscopic study of 3000 yar-old human skeletal remains from a sambaqui, Santa Catarina, Brazil. *J Raman Spectrosc*. 32, p. 17-22.

FILIPPINI, J. (2004) – *Biodistância entre sambaquieiros fluviais e costeiros: Uma abordagem não-métrica craniana entre três sítios fluviais do vale do Ribeira – SP (Moraes, Capelinha e Pavão XVI) e três costeiros do sul e sudeste do Brasil (Piaçaguera, Jabuticabeira II e Tenório*. MSc thesis, Universidade de São Paulo, Brazil.

FILIPPINI, J.; EGGERS, S. (2005/2006) – Distância biológica entre sambaquieiros fluviais (Moraes-Vale do Ribeira- SP) e construtores de sítios litorâneos (Piaçaguera e Tenório – SP e Jabuticabeira II-SC). *Revista do Museu de Arqueologia e Etnologia* USP 15-16:165-180.

GARRIGAN, D.; HAMMER, M.F. (2006) – Reconstructing human origins in the genomic era. *Nat Rev Genet*. 7:9, p. 669-680.

GASPAR, M. D. (1998) – Considerations of the Sambaquis of the Brazilian Coast. *Antiquity*. 72:277, p. 592-613.

GUATTELLI-STEINBERG, D.; IRISH, J. D.; LUKACS, J. R. (2001) – Canary islands-north African population affinities: measures of divergence based on dental morphology. *Homo*. 52, p. 173-188.

HAUSER, G.; DE STEFANO, G.F. (1989) – *Epigenetic Variants of the Human Skull*. Stuttgart: Schweizenberger Verlag.

HIERNAUX, J. (1963) – Heredity and environment-their influence on human morphology. A comparison of two independent lines of study. *Am J Phys Anthrop*. 21: 4, p. 575-589.

IRISH, J.D. (2000) – The Iberomaurusian enigma: North African progenitor or dead end? *J Hum Evol*.. 39, p. 393-410.

IRISH, J.D.; GUATTELLI-STEINBERG, D. (2003) – Ancient teeth and modern human origins: an expanded comparison of African Plio-Pleistocene and recent world dental samples. *J Hum Evol*. 45, p. 113-144.

LARSEN, C.S. (1997) – *Bioarchaeology: Interpreting behavior from the human skeleton*. Cambridge: Cambridge University Press.

NICHOL, C.R. (1989) – Complex segregation analysis of dental morphological variants. *Am J Phys Anthropol*. 78, p. 37-59.

OKUMURA, M.M.M. (2008) – Diversidade morfológica craniana: micro-evolução e ocupação pré-histórica da costa brasileira. *Pesquisas* (UNISINOS): 66:1-306.

SCOTT, G.R. (1973) – *Dental morphology: a genetic study of American white families and variation in living southwest Indians*. Ph.D. dissertation. Arizona State University, Tempe, USA.

SCOTT, G.R.; TURNER II, C.G. (1997) – *The anthropology of modern human teeth. dental anthropology and its variation in recent human populations*. Cambridge: Cambridge University Press.

SJØVOLD, T. (1977) – Non-metrical Divergence Between Skeletal Populations; the Theoretical Foundation and Biological Importance of C.A.B. Smith's Mean Measure of Divergence. *OSSA International Journal of Skeletal Research*. 4:1. Solna: Osteological Research Laboratory, University of Stockholm.

SMITH, C.A. (1977) – A note on genetic distance. *Ann Hum Genet*. 40, p. 463-479.

SPARKS, C.S.; JANTZ, R.L. (2002) – A reassessment of human cranial plasticity: Boas revisited. *Proceedings of the National Academy of Sciences USA*. 99:23, p.14636-14639.

TURNER II, C.G.; NICHOL, C.R.; SCOTT, G.R. (1991) – Scoring procedures for key morphological traits of the permanent dentition: The Arizona State University Dental Anthropology System. In Kelley, M.; Larsen, C., eds, – *Advances in Dental Anthropology*. New York: Wiley Liss, p. 13-31.

MICROFOSSILS IN DENTAL CALCULUS FROM A BRAZILIAN SHELLMOUND: WHERE DID THEY COME FROM?

Célia Helena BOYADJIAN, Sabine EGGERS
Laboratório de Antropologia Biológica, Departamento de Genética e Biologia Evolutiva,
Instituto de Biociências da USP, Rua do Matão, 277, Cep: 05508-900 São Paulo, Brazil
e-mail:saeggers@usp.br
fone: 3091-7588, fax: 3091-7553

Karl REINHARD
School of Natural Resources, University of Nebraska-Lincoln, Lincoln, Nebraska
Sponsor: Fundação de Amparo a Pesquisa do Estado de São Paulo (Fapesp – Grant 05/51710-1 and 04/11038-0)

***Abstract**: The enormous amounts of mollusk shells that compose the Brazilian shellmounds (Sambaquis) were thought to be food remains. However, zooarchaeological and stable isotope studies showed that fishing was the basis of the shellmound builder's subsistence. Although plants do hardly preserve in these shellmounds, anthracological analyses revealed that they played an important role in daily activities. Fortunately, eating plants and manufacturing artifacts with the help of teeth leave microfossils, like starch grains, phytoliths and fibers, trapped in the dental calculus matrix. Their identification and quantification permit reconstructing diet and the use of plants in the past. In the present paper we analyze the dental calculus of the Sambaqui Jabuticabeira II at the Southern Coast of Brazil. Analyzing the relative proportions of plant microremains in dental calculus from individuals buried in this site unravels questions about the plants consumed and used in manufacture and, eventually, also the way they were prepared. We also present here preliminary results on a technique of microfossil recovery from cryptic dental calculus obtained from the fluvial shellmound Moraes, located in the inland, in order to compare the results with Jabuticabeira II. We found many different microstructures in the dental calculus, mainly starch, including some modified grains. Several microfossils still await further identification. Nevertheless, it is possible to state that the starch grain concentrations from Moraes´ calculus are much higher than that from Jabuticabeira II. This parallels the findings that there is a significantly higher caries rate in Moraes than in Jabuticabeira II. This indicates differences in the diet, albeit the results presented here are still preliminary. More samples will be processed and new test will be carried out to test whether this holds true.*
***Key words**: Shellmounds, Brazil, dental calculus, plant microfossils, starch grains*

INTRODUCTION

Shellmounds are distributed along almost the entire coast of Brazil. These archeological sites are composed mainly of mollusk shells but can contain faunal remains, lithic and bone artifacts, habitation structures and also burials.

For a long time, it was believed that the enormous amounts of shells were food remains. However, zooarchaeological and stable isotope studies showed that fishing was the basis of the shellmound builder's subsistence. The shells were largely used as construction material (Bandeira, 1992; Figuti, 1993; de Masi, 1999; Klökler, 2001).

Nevertheless, recent studies show that plants also played an important role in their daily activities (Scheel-Ybert 2000; 2001; Scheel-Ybert *et al.*, 2003; Reinhard *et al*, 2001). But still little is known about the vegetal consumption of these groups, because plant macroremains are very poorly preserved in this type of site (Scheel-Ybert *et al.*, 2003).

Fortunately, eating plants and manufacturing artifacts such as ropes and baskets with the help of teeth leave microfossils trapped in the dental calculus matrix. As the dental calculus (tartar) is formed, some micro remains like starch grains, phytoliths and fibers originating from the plants chewed can get trapped in its structure (Fox *et al*, 1996). These microfossils become preserved for thousands of years, because of the calcium carbonate matrix of the calculus. Their identification and quantification permit reconstructing diet and the use of plants in the past.

Anthracological studies show evidences of the use of different wood types, edible seeds and tubers at various shellmound sites in Brazil (Scheel-Ybert 2000; 2001; Scheel-Ybert *et al.*, 2003). One of those is the huge cemetery Jabuticabeira II, located at the southern coast. The caries frequency in this site, however, is very low (0.44% affected teeth, Okumura & Eggers, 2005), suggesting a diet low in cariogenic food. More recently, skeletons have been excavated from the layer of black earth that covers part of the site. This black earth is seen by some archaeologist as an indication of incipient horticulture. If this hypothesis is true, the caries frequency in the skeletons recovered from this area should be higher.

These facts constitute the basis for the present study of the microfossils in dental calculus of the Jabuticabeira II shellmound. Analyzing the relative proportions of plant microremains in dental calculus from individuals buried in the shell as well as in the dark earth layers will give us clues about the plants consumed, used in manufacture and, eventually, also the way they were prepared. Starch grains, for example, if exposed to heat, change their three-dimensional structure and the modified starch can be detected by light microscope (Cortella & Pochetino, 1994).

In this article we show the preliminary results of our research on the dental calculus contents of individuals buried at Jabuticabeira II shellmound. The data from the black earth sediments have not been obtained yet. We also present preliminary results on a new technique of microfossil recovery from cryptic dental calculus obtained from the fluvial shellmound Moraes, located at the southeastern region of Brazil (about 40km from the seashore), in order to compare the results with Jabuticabeira II.

MATERIALS AND METHODS

To recover the microfossils from Jabuticabeira's dental calculus, we used the method described by Reinhard *et al* (2001). First, we detached fragments of dental calculus from the teeth of 19 adult individuals excavated from the shell layers. Six of them are male, 5 female and 8 of undetermined sex.

Since many of the individuals from Jabuticabeira II have thick deposits of dental calculus in many teeth, we could obtain enough samples using only few teeth per person. This also had the advantage to leave teeth with calculus deposits for future research. It is important to say that, in some cases, because of the large amount of the deposits, it was possible to obtain 2 samples for the same person; one from anterior teeth and one from posterior teeth.

The dental calculus fragments were then processed in a solution of 5% hydrochloric acid. We put the calculus fragments in labeled microtubes, filled them with hydrochloric acid and added a lycopodium spore tablet to allow the quantification of microfossils (Pearsall, 2000). After a few minutes, when the lycopodium tablets were already dissolved, we centrifuged the tubes 3 times. The first time we used hydrochloric acid solution but for the other 2 times we substituted the acid by distilled water. Then, we decanted the water, rinsed the sediments out of the tubes with ethanol and transferred the pellet to a labeled vial with a lid. Each vial contained one sample.

Five slides with 10μL of each sample were prepared and then scanned systematically under a light microscope with 20x and 40x magnification, in order to classify and count the microfossils. Then we could estimate the microfossils concentration of each sample using a pollen concentration formula, as reviewed by Reinhard *et al.* (2005).

In contrast to the teeth excavated in Jabuticabeira II, the individuals from Moraes did not show thick deposits of dental calculus. In the contrary, the calculus is only a cryptic and very thin deposit, which is impossible to be detached from the teeth mechanically. This constitutes a methodological challenge. We have developed and preliminarily tested a new technique, called dental wash, that allows the recovery of microfossils from these faint dental carlculus marks (Boyadjian *et al.*, 2007).

The samples from Moraes were obtained with this new method, but the analysis under light microscope, the identification and quantification of microfossils of these samples were carried out in the same way as described for Jabuticabeira II.

RESULTS AND DISCUSSION

We found many different microstructures in the samples. A part of them are not yet identified. So, in this paper we will concentrate on the more detailed description and discussion of the starch grains.

The concentration of starch grains for each of the samples can be seen in Figure 5.1A. In some samples we could not find any starch grain and in the majority of the cases (18 samples or 15 individuals) the starch grain concentration is lower than 10grains/μL.

When we separate the samples by sex we note that, with the exception of one sample, the highest concentrations of starch can be found among men (Figure 5.1B).

Separating the samples by age (Figure 5.1C), we note that the highest starch grain concentrations can be observed in the group of middle adults. This is expected because the dental calculus accumulates progressively, so, the deposits grow with age (Ânerud *et al.*, 1991). Possibly the amount of microfossils trapped in it also gets higher. Too few old adults were available to test this trend.

The starch grains found measure 5μm up to 25μm in diameter. Their shape also varies. We found spherical (figure 5.2A and B), ovoid, bell-shaped (figure 5.2C) and polygonal grains (figure 5.2D and E).

Some of the grains showed fissures from the hilum (figure 5.2F). The hilum is a diagnostic feature of a starch grain (Loy, 1994 apud Pearsall, 2000), but in some cases it was not visible (figure 5.2G). All of the grains showed the extinction cross under polarized light, but in some cases it was not very well defined (figure 5.2D). As the hilum, the extinction cross is also a diagnostic feature of the starch grain (Loy, 1994 apud Pearsall, 2000). In just one sample we found some compound grains (figure 5.2H).

There is also some evidence of starch grains with a differrent, modified three-dimensional structure. This taphonomic changes or that these starch grains were consumed after cooking, because heat modifies starch grain structure.

Experimentally applying the new dental wash method (Boyadjian *et al.*, 2007) on 3 males, 5 women and 16 individuals of undetermined sex from Moraes we have also found starch grains. Curiously these starch grains look different in shape than those from JabuticabeiraII.

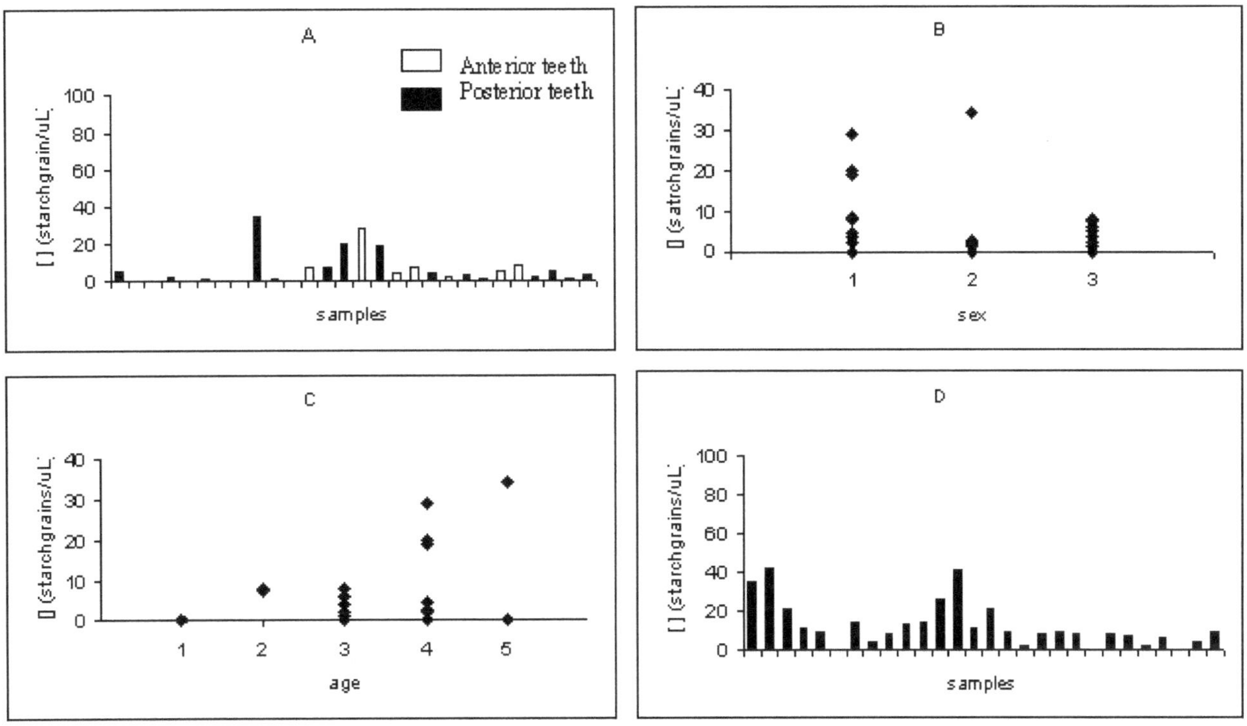

Fig. 5.1. Starch grains concentration in: A) Jabuticabeira II; B) Jabuticabeira II in samples separated by sex (1=males, 2=females, 3=unknown); C) Jabuticabeira II in samples separated by age at death (1=child, 2=juvenil, 3=young adult, 4=middle adult, 5=old adult); and D) in Moraes

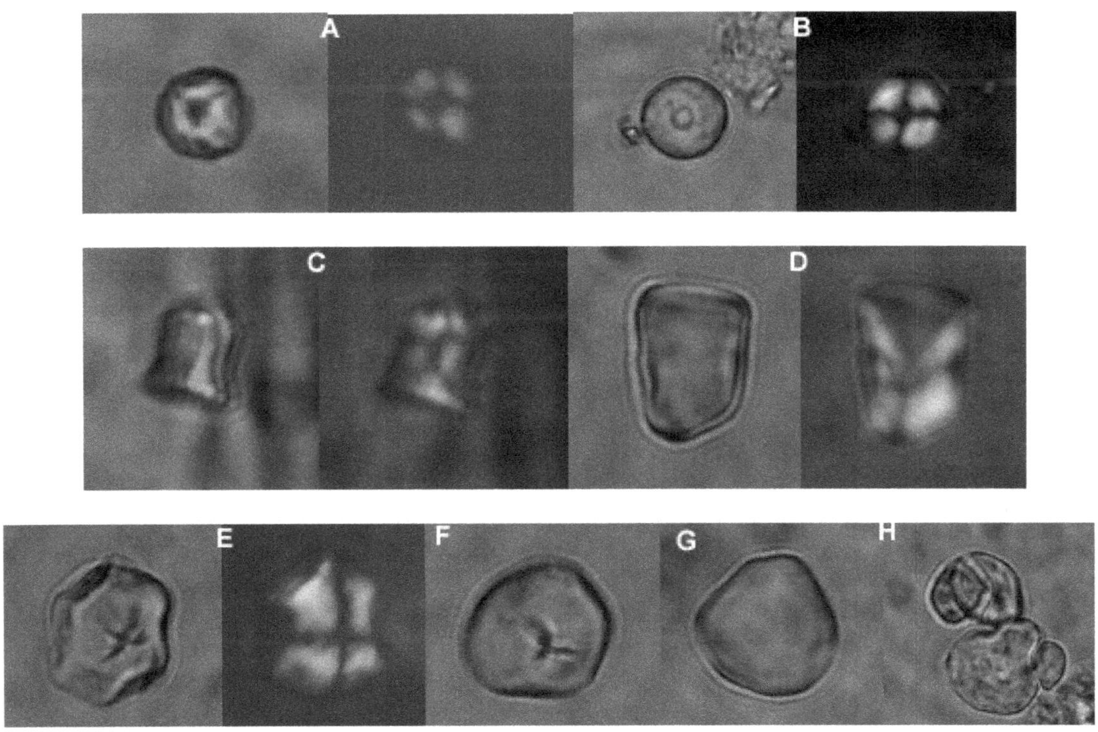

Fig. 5.2. Examples of starch grains found in the dental calculus from Jabuticabeira II. Where there are two pictures of the same grain, the dark one was taken under the polarized light to show the extinction cross. All the pictures were taken under 40x magnification

Besides that, the concentration of starch grains in the samples of Moraes is significantly higher (t=-2,05, p=0,04) than in the samples of Jabuticabeira II, as is shown in figure 5.1 (A and D).

In parallel to this finding we also found a significantly higher frequency of dental caries in Moraes (12%) as opposed to Jabuticabeira II (0.44%) individuals. These results point to a more important ingestion of cariogenic food by the individuals form Moraes. This is in accordance with the higher concentration of starch grains found in dental calculus from Moraes than in Jabuticabeira II.

However, the new technique used in Moraes still needs to be improved. Additionally, we will analyze the sediment surrounding teeth and other parts of the skeleton in order to test if these results are influenced by contamination.

Besides starch grains, we also found phytoliths (silica or calcium oxalate bodies) in the samples of Moraes. In Jabuticabeira II, as stated before, we found many structures not yet identified, but we also observed cells from vascular plant tissue. These are typical structures from plant organs such as stems. So, these findings indicate that the individuals from Jabuticabeira II used plant stems as food and/or the stems could have been manipulated with the teeth for the production of ropes or baskets.

CONCLUSIONS

Microfossils, such as starch grains, phytoliths and other plant structures become preserved for thousands of years in dental calculus. Their identification and quantification permit reconstructing diet and the use of plants in the past. It is not yet possible to distinguish between collected and cultivated plants. However, we found modified starch grains, which suggest food preparation (such as cooking or roasting).

In the coastal shellmound Jabuticabeira II we found many different structures in the dental calculus, mainly starch grains, but there are several microfossils that still need to be identified.

Comparing Jabuticabeira II to the inland shellmound Moraes, we noted that the starch grain concentrations from Moraes´ calculus are much higher than that from Jabuticabeira II. This parallels the findings that there is a significantly higher caries rate in Moraes than in Jabuticabeira II, which could indicate differences in the diet.

The results presented here are preliminary. More tests for the new technique called dental wash applied in Moraes will be carried out, more samples from Jabuticabeira II, especially from the black earth sediment, will be processed and analyzed.

Acknowledgement

We would like to thank the Fullbright Comission's grant to Sabine Eggers for supporting Professor Karl Reinhard´s visit to Universidade de São Paulo, Brazil. We also thank Paulo De Blasis and Levy Figuti, from Museu de Arqueologia e Etnologia da Universidade de São Paulo (MAE-USP), responsible for the archaeological projects which permitted the burials to be analyzed; Maria Luiza Faria Salatino from Laboratório de Fitoquímica do Departamento de Botânica do Instituto de Biociências, Universidade de São Paulo, who generously gave access to her laboratory and equipment for the final steps of the experiment. The research presented here was supported by FAPESP (Grants 05/51710-1 and 04/11038-0).

References

ÅNERUD, Å.; LÖE, H.; BOYSEN, H. (1991) – The Natural History and Clinical Course of Calculus Formation in Man. *Journal of clinical Periodontology* 18 p160-170.

BANDEIRA, D.R. (1992) – Mudança na Estratégia de Subsistência do Sítio Arqueológico Enseada I: Um estudo de caso. Tese de Mestrado, Universidade de São Paulo, Brazil. 127 p.

BOYADJIAN, C.H.; EGGERS, S.; REINHARD, K. – (2007) Dental wash: a new method of estimating microfossil content in prehistoric teeth. *Journal of Archaeological Sciences* 34:1622-28.

CORTELLA, A.R.; POCHETTINO, M.L. (1994) – Starch Grain Analysis as a Microscopic Diagnostic Feature in the Identification of Plant Material. *Economic Botany* 48:2, p171-181.

De MASI, M.A.N. (1999) – Prehistoric hunter-gatherer mobility on the southern Brazilian coast: Santa Catarina Island. Tese de Doutorado, Stanford University. USA. 186 p.

FIGUTI, L. (1993) – O Homem Pré-histórico, o Molusco e os Sambaquis: considerações sobre a subsistência dos povos sambaquieiros. *Revista do Museu de Arqueologia e Etnologia* 3 p. 67-80.

FOX, C.L.; JUAN, J.; ALBERT, R. (1996) – Phytolith Analysis on Dental Calculus, Enamel Surface and Burial Soil: Information about Diet and Paleoenvironmet. *American Journal of Physical Anthropology*, 101 p.101-114.

KLÖKLER, D.M. (2001) – Construindo ou deixando um sambaqui?: análise de sedimentos de um sambaqui do litoral meridional brasileiro – processos formativos – Região de Laguna-SC, MS Dissertation, Universidade de São Paulo, Brazil.

LOY, T.H. (1994) – Methods in the Analysis of Starch Residues on Prehistoric Stone Tools. In *Tropical Archaeobotany: Applications and New Developments*, Edited by Jon G. Hather, p 86-114. Routledge, London. Appud Pearsall, D.M. (2000). Paleoethnobotany: A Handbook of Procedures. New York: Academic Press. Second Edition.

OKUMURA, M.M.M.; EGGERS, S. (2005) – The People of Jabuticabeira II: Reconstruction of the Way of Life in a Brazilian Shellmound. International Journal of Osteoarchaeology. *Homo*. 55 p.263-281.

PEARSALL, D.M. (2000) – Paleoethnobotany: A Handbook of Procedures. New York: Academic Press. Second Edition.

REINHARD, K. [*et al.*] (2001) – Microfossils in Dental Calculus: A New Perspective on Diet and Dental Disease. In Human Remains: Conservation, Retrieval and Analysis. Proceedings of a conference held in Williamsburg, VA, Nov 7-11th 1999. Edited by Emily Williams.

REINHARD, K.J.; EDWARDS, S.K.; MEIER, D.K. (2005) – Pollen concentration analysis of Salmon Ruin and Antelope House: Documenting Anasazi Dietary Variation, *Journal of Palaeogeography, Palaeoclimatology, and Palaeoecology* 237 p. 92-109.

SCHEEL-YBERT, R. (2000) – Vegetation stability in the Southeastern Brazilian coastal area from 5500 to 1400 ^{14}C yr BP deduced from charcoal analysis. *Review of Palaeobotany and Palynology* 110 p. 111-138.

SCHEEL-YBERT, R. (2001) – Man and vegetation in the Southeastern Brazil during the Late Holocene. *Journal of Archaeological Science* 28:5, p. 471-80.

SCHEEL-YBERT, R. [*et al.*] (2003) – Novas perspectivas na reconstituição do modo de vida dos sambaquieiros: uma abordagem multidisciplinar. *Revista de Arqueologia* 16 p109-137.

TEETH, NUTRITION, ANEMIA, INFECTION, MORTALITY: COSTS OF LIFESTYLE AT THE COASTAL BRAZILIAN SAMBAQUIS

Sheila MENDONÇA de SOUZA, Veronica WESOLOWSKI
Escola Nacional de Saúde Pública / FIOCRUZ (sferraz@ensp.fiocruz.br)

Claudia RODRIGUES-CARVALHO
Museu Nacional, Universidade Federal do Rio de Janeiro

Abstract: *The bioarchaeology of Brazilian sambaqui peoples has been improved in the last 30 years. An increasing number of papers is published and new data about nutrition, mobility, health, taphonomy and cultural data are produced. Dental pathology and othe information about health is abundant, and subjects such as infections, stress markers, anemia among other are being explored. Paleodemography still depends on the future bigger excavations providing better samples. Difference in lifestyle, along the coast, is suggested by different pathocenosis. In general, anemia is very frequent, chronic infections are endemic and caries rates are low, while tooth abrasion is intense. Stress indicators (LEH and Harris lines) are also frequent, suggesting that in spite of a diet rich in protein, there was physiologic stress following the weaning and infancy. New data about the use of carbohydrates in the diet start to elucidate special cases with high prevalence of dental caries.*
Key words: *Shell mounds, paleoepidemiology, South America, Prehistory, bioarchaeology*

Resumé: *Les derniers 30 ans virent un vrai avance à la recherche bio-archéologique aux sambaquis brésiliens et une grande quantité d'articles fut publiée. Même si la pathologie dentaire reste le sujet plus recherché, il y a aussi articles publies aux infections, aux signes d'estress, a l'anémie, etc. Pour la paleodemographie il faut encore former des ensembles funeraires plus nombreuses et representatives. Si la pathocenose de ces groupes signale les diferences d'style de vie au long du litoral, il y a un cadre general caractérisé pour l'endemicité des infections croniques, l'haute frequence de l'anémie et basse prevalence de carie. Les hypoplasies d'email dentaire et les lignes de Harris sont aussi frequentes et suggèrent que la diète hyperproteic n'empechait pas l'existence d'stress physiologique pendant l'infance et l'adolescence dans cettes populations.*
Mots clés: *Amas de coquilles, paleoepidemiologie, Amerique du Sud, Prehistoire, bioarchaeologie*

INTRODUCTION

Skeletal morphology points to health and disease among other bioarchaeological information, helping to reconstruct the lifestyle of past populations (Larsen, 1999). The interest for understanding life conditions has been pointed out decades ago (Hooton, 1930), and since them paleopathology is also paleoepidemiology (Angel, 1966; Buikstra & Cook, 1980; Cohen & Armelagos, 1984; Souza, Carvalho & Lessa, 2003, among others). Skeletal and teeth stress indicators as well as disease indicators in their proper context provide hypothesis, biocultural models (Goodman *et al.*, 1988; Souza, 1999) and Health Indexes (Steckel, Sciulli & Rose, 2002). Bioarchaeology of *sambaquis* has been using big and also small numbers in order to explain how prehistoric coastal populations lived and died. In Brazil more and more papers about bioarchaeology of sambaquis are being published in the last 30 years.

Until the seventies there was no systematic information about paleopathology of the *sambaqui* people, except for some data on dental health (Cunha, 1963a, 1963b; Ferraz & Mendonça de Souza, 1973). Since them a lot of information was provided on paleopathology. Anemia, dental loss, dental caries, dental calculus, periosteal reactions on bones and infection are the most popular studies (Machado & Kneip, 1994, Souza, 1999; Storto, Eggers & Lahr, 1999, 2001; Wesolowski, 2000; Souza, Wesolowsky, Rodrigues-Carvalho & Lessa, 2006). Trauma and workload came to be subject of research very recently (Lessa & Medeiros, 2001; Rodrigues-Carvalho, 2004). Thanks to this refreshing discussion paleopathology of coastal populations is now more integrated to the archaeological models, opening new windows to the understanding of the past. The studies concerning food remains, isotopic analysis of bones, demography and others are still at the beginning. New insights help testing hypothesis about the adaptative cost of different lifestyles along the shore line.

The present literature survey includes 119 papers, 56 sites (21 from Rio de Janeiro State, 12 from São Paulo State, 4 from Paraná State, 19 from Santa Catarina State) and a total amount of 1546 skeletons. To this was added some unpublished data. Not all the bibliographic references are cited here but the data summarizes the main results of the revision. An increasing amount of information on paleopathology, more comparisons between sites, and new insights and hypothesis about prehistory as well as lifestyle and health is now available. Despite the overall similarities among the *sambaquis* it is already possible to confirm that those coastal groups had different life conditions, depending on time, ecology and culture. Sharing many aspects of the environment, subsistence strategies and material culture, they also had substantial differences in diet, physical activity and settlement. Cultural differences are not very well known though it can actually explains part of the diversity pointed by paleopathology.

The aim of the present paper is to bring general information about biological variability of Brazilian *sambaqui* people under the point of view of dental pathology, nutrition, infection and mortality.

WHAT DO WE KNOW ABOUT *SAMBAQUI* PEOPLE?

Sambaqui people were considered well adapted, well nourished, robust and living on a diet rich in protein (fish and also shellfish, fruits/roots/seed collected along the *restingas*, swamps and Atlantic forest along the coast). No caries and low rates of teeth loss are described for the main sites, reinforcing the idea of good health conditions. Their low to medium stature was not quite different from the stature of other groups in Brazil. Their very thick bones were suggestive of well calcified and developed skeletons. Their long bones marked by muscular insertions suggested strong complexion. For decades, Physical Anthropologists considered the *sambaqui* people the very model of a well adapted, strong and healthy people living in an affluent environment for millennia.

When specialists in paleopathology finally started to dedicate themselves to the study of those groups, new insights came to give more precise idea of their varied health and lifestyle. Perhaps not all those groups had good health, or perhaps we should review the concepts and methods in paleopathology. Perhaps the adaptability along the time and space in the coastal areas is such as to create a great variety of life conditions. At this very moment data show us a rich bioarchaeological scenario to be investigated.

TEETH

The first paleopathology studies on *sambaqui* people were focused on teeth. Between the fifties and the seventies Ernesto de Mello Salles Cunha (Cunha, 1963a; 1963b) published surprising low rates of caries in *sambaqui* series, including some cases of caries absence, suggesting they could had naturally resistant teeth for genetic or dietary reasons. Their low dental loss during life, the expressive dental calculus and dental wear were also considered typical of those harvesting people. More recent papers definitely proved a very unequal picture concerning teeth decay, suggesting that not all the coastal *sambaqui* people had the same pattern of oral health (Wesolowski, 2000; Haubert *et al.*, 2004).

The three main indicators in published and unpublished data are caries, tooth wear and "in vivo" dental loss. Caries prevalence *per* person vary from 0% to 2% in sites such as Cabeçuda, Rio Vermelho, Laranjeiras I, Cabeça do Índio, Forte Marechal Luz and Jaboticabeira II, to 43-60% in sites such as Enseada I and Rio Comprido. A few sites have high prevalence of caries (Neves, Unger & Scaramuzza, 1984). The explanation to this fact is not clear yet, but the first studies proved that caries are definitely not associated to the ceramics that occur in a few *sambaquis*, at least in Santa Catarina State (Wesolowski, 2000). In Rio de Janeiro *sambaquis* have a medium prevalence of caries, for instance, 22% of caries per teeth in Beirada site (Machado & Kneip, 1994).

The alveolar abcesses are highly prevalent (40-70%) even in the absence of caries, possibly associated to intense tooth wear and periodontal trauma. The non mastigatory wear caused peculiar spiral and beveled forms in teeth also occursed in specific sites, possibly caused by the processing of some raw material. It is described in the sites of Jaboticabeira, Forte Marechal Luz, Cabeçuda (Santa Catarina State), Piaçaguera and Tenório (São Paulo State) and Beirada, Embratel, Zé Espinho (Rio de Janeiro State). Dental loss during life is rare, perhaps because of the sticking effect of hypercementosis. Dental calculus varies in prevalence and intensity, and the study of micro-fossils in calculus (Reinhard *et al.*, 2001) is confirming different prevalences of starch, phytoliths and processed parts of plants such as fibers, epidermis and spikes. In the Cabeçuda *sambaqui* tooth loss of the incisors in males (17%) was associated to trauma and possible use of lip ornament (Rodrigues-Carvalho & Souza, 1998).

Information about periodontal disease is scarce, except for the report of apical cavities generally not associated to dental loss or caries. Masticatory trauma caused by bones, minerals, fragments of shells and possibly plant remains such as phytoliths fit well the explanation of infection and cavities.

Mandibular and maxillary *tori* are also usually frequent at the skulls. Although they could be explained by mechanical as well as infectious stimulus (Alvim & Soares, 1984; Souza, 1999) most of the Brazilian literature still considers them as epigenetic traits. Auditive *tori* are also frequent and their interpretation is that of an epigenetic trait, as well as the mandibular and the maxillary *tori* (Alvim & Soares, 1984; Alvim, Soares & Cunha, 1984). An alternative etiologic explanation is that they could be associated to low temperatures of the water – diving – or the air – exposure to strong cold winds. A periosteal response to local infection can be an alternative explanation (Souza, 1999), although Peixoto (1989) in his study of Cabeçuda skulls proposed that the position of the *tori* in the inner part of the ear was suggestive of a genetic etiology. The actual prevalence of external auditive *tori* in Santa Catarina State collections is between 22% and 36% of the individuals, and could be explained either by cold waters or winds (Eggers *et al.*, 2006). In Rio de Janeiro and São Paulo *sambaquis*, where the environmental conditions are less severe and the waters less cold there is almost no *tori*, except in Condomínio do Atalaia, a site placed at the only point of frozen resurgent deep sea waters along the Rio de Janeiro coast.

Comparing *sambaqui* people to modern people, we realize that along the Brazilian coast low carie rates, intense wear and medium to high periodontal reaction have also been observed among the traditional fishermen communities (Pourchet & Alvim, 1975), who share some aspects of their environment and lifestyle with the *sambaqui* people.

ANEMIA

During the eighties the Brazilian *sambaquis* started to be studied also for anemia, with the help of indirect indicators such as porotic hyperostosis (PO) and *cribra orbitalia* (CO). The first description of high prevalence of PH and CO in a Brazilian skeletal series was published by Alvim & Gomes (1989), this study was followed by others. Data to some of the biggest *sambaqui* collections like Piaçaguera and Cabeçuda, is today available. PH and CO were found in high prevalence in *sambaqui* skeletons, suggesting that anemia was a real problem in infancy, only surpassed after the 10th year of life. Once the diet of those prehistoric foragers did not lack iron, only the imbalance between iron intake and consumption would explain the data. Parasitism, infections, diarrhea and other factors could be associated to iron loss. Iron consumption by parasites, iron loss associated to blood loss, or adaptative response to infections during childhood (Alvim & Gomes, 1989; Souza, 1999) can be alternative explanations to anemia in such a well nourished people.

The first interpretation of anemia in Brazilian s*ambaqui* people also considered that the accumulation of such a discarded pile of carcasses at the settlement induced a poor sanitation area. According to Alvim & Gomes (1989), in a tropical environment such a condition would cause people to be exposed to high infestation risks. This interpretation proposed that anemia could be a direct consequence of massive parasitism, although no coprolites from sambaquis could ever been analyzed. In the nineties, an alternative hypothesis was proposed by Souza (1999). Following Stuart-Macadam & Kent (1992) model it was suggested that anemia could be just a consequence of the daily exposition to infections, especially salt water bacteria of the genus *Vibrio,* universally present in the coastal environments. Objects, materials, food and body surfaces in contact with salt water are generally colonized by that *Vibrio,* especially during the warm season, potentially causing gastroenteritis and other infections of the ear, eyes and skin, as described for modern populations living by the sea or using sea food (Daniels *et al*, 2000). A permanent infectious stimulus from the early infancy, universal and constant, could explain anemia as an adaptative response in children living a *sambaqui* lifestyle.

Piles of discarded remains or intentionally built, as suggested by Gaspar (2003) and other archaeologists, the lifestyle associated to the *sambaquis* would be consistent with the permanent exposition to bacteria. In such a condition it is possible to assume an adaptative response of iron consumption and sustained low level in the blood, as proposed by Stuart-Macadam & Kent (1992). Differences in the prevalence of anemia in the sites possibly point to different environmental exposition and risks, and that is consistent changing lifestyles along the coastal environments.

Besides the salt water *Vibrio,* other pathogens are proposed to have been present among Brazilian *sambaquis* groups. In fact, under the paleoepidemiologic point of view data point to the fact that endemic infections spread out among the *sambaqui* people in different places and periods, bones are affected in the form of periosteal reactions, occasional osteítis and osteomielitis, as discussed ahead, what could also explains anemias and other stress indicators.

NUTRITION

For a long time nutrition has been considered good among *sambaqui* people. Supposed to live affluent lives, the coastal groups were supposed to feed on coastal environments. Diet was supposed to be hyperproteic, based on fish and shellfish, and this was generally associated to good health. Although their stature was medium to low (Alvim, Vieira & Cheiuche, 1975) their skeletons were robust and marked dimorphic, suggesting a good development of the genetic potential. Added to low rates of dental decay, this general description to *sambaquis* people was supposed to point to good health (Souza, 1995). Variation in *sambaquis* series was not discussed except for one comparative study of skeletal series suggesting more than one morphological type (Alvim, 1978). On the other hand, the archaeologists agreed that the *sambaquis* expressed different cultural manifestations in different places and times (Prous, 1992) leading to think about different feeding strategies. Lima (1991) proposed that the differences among the sites could express different access to resources, competition and some level of complexity in those societies. Hierarchy between sites along the same coastal area has been suggested by Gaspar (2003) and Lima (1991). Living at the coast was supposed to be living in an affluent environment, and *sambaqui* people were not hypothesized as people having big health problems. Natural risks of their lifestyle were not discussed until the nineties (Souza, 1995; 1999).

Nutritional stress was also suggested by the finding of expressive linear enamel hypoplasia. The first results indicating high prevalence of that stress indicator among the *sambaqui* people was also a surprise (Souza, 1999). According to Goodman & Capasso (1992) a low carbohydrate intake even associated to a hyperproteic diet could have physiologic impact and produce considerable physiologic stress. This possibility of low carbohydrate intake unbalancing diet was first considered to explain the prevalence of hypoplasias in *sambaquis* for the site of Cabeçuda (Souza, 1999). Starch in dental calculus (Reinhard *et al*., 2001; Wesolowski *et al*., 2007), isotope studies in bones (De Masi, 1999) and also charcoal analysis (Scheel, Gaspar & Ybert, 1996) will certainly be helpful in the future to give us a full picture of health and disease. The enamel defects are generally formed between 2 and 4 years in coincidence with the weaning period. Otherwise, in Enseada I the stress causing enamel defects comes later, between 4 and 6 years of age. But indeed, polishing of the enamel surfaces, specialy in those samples with non masticatory wear, makes much more difficult to compare the results among different sites. More detailed microscopic analysis is needed to improve the results.

In Cabeçuda *sambaqui* 64% of the incisors have LEH (Souza, 1999). In Rio Vermelho, with a diet based on fish, 53% of the women and 26% of the men have linear enamel hypoplasias (Hube, 2005). In Morro do Ouro LEH rate is 33-42%, in Enseada I is 51%, both are sites with high prevalence of caries and possibly higher carbohydrate intake. Both the archaeological analysis of animal remains (Figuti, 1993), and the isotopic analysis of bones (de Masi, 1999), point to fish as the main protein source, contrary to the former interpretation of mollusks being the staple food.

Perhaps diet was more balanced in the sites where starch plants were used. Studies on isotopes, as well as investigation of microfossils from teeth, now in curse suggest the differences in diet among the sites can be greater than previously supposed. The more we know the more we think about the differences in nutrition, anemia and adaptation in the Brazilian sambaquis.

INFECTION

Infection is proved to be very frequent among the *sambaquis* people, especially periosteal reactions. Although most of the papers reporting periostitis do not distinguish wich lesions could be traumatic, the infectious etiology can be barely inferred by the report of systemic, symmetrical and/or poliostotic lesions. The tropical climate, the concentration of people and groups, and the exposure to especific coastal risks such as salt water vibrios make the infections etiology – either specific, or inespecific – a strong epidemiological hypothesis (Souza, 1995, 1999). Considering the prevalence of infections, including systemic/poliostotic ones can help understanding if not all the infections, the main endemic ones. Focal periosteal reactions are also described for in sambaquis squeletons. Their possible infectious etiology can be explained by hematological dissemination, or contiguous dissemination from skin for instance. Excluding the focal periostitis as possibly traumatic can cause underestimation of the prevalence of infections, excluding occasional, minor focal lesions.

Perisotitis, osteítes and osteomielitis are described in the *sambaqui* skeletons, suggesting that inespecific as well as specific infections affected the bones. Periosteal reactions, either laminar or fibrous, and chronic periostitis associated to osteitis affect the long bones, mainly the lower limbs. At least one case of osteomyelitis is present in each site. Reactions at the outer table, vault scars and porosity are also describe in rare cases. Focal periosteal reactions in maxilla and mandibular, are generally associated to dental foci.

In Rio de Janeiro State, Messias (1977) described, "discrete hyperostosis" in tibias from Sambaqui do Forte. Rodrigues *et al.* (2001) mentioned one individual with osteomielitis. Souza (1995) describes an impressive case of chronic poliostotic osteomielitis in a skeleton from the Arapuã, where other individuals with signs of infection have been identified. Infections are also mentioned for the Condominio do Atalaia. In Zé Espinho *sambaqui* about 60% of the skeletons have periostitis, one of them has disseminated lesions including osteitis. In Ilhote do Leste 90% of the skeletons also have periostitis, half of them are systemic lesions. In Beirada, Boa Vista and Algodão, 50 to 80% of the individuals have periosteal reactions suggestive of infection, as in other sites, at least one skeleton of each site has more disseminated or severe lesions. Most of the skeletons described for periostitis are adults, the immature not being mentioned. In São Paulo there is no systematic study, but initial descriptions by Silva (2005) mention skull and ribs periosteal reactions in Buracão and Tenório.

In Santa Catarina 60% of the skeletons from Jaboticabeira have periostitis/ osteitis/osteomielytis, 47% of them are severe (Storto, Eggers & Lahr, 1999, 2001) In Cabeçuda are symmetrical lesions in tibias (Souza, 1995). In Forte Marechal Luz periosteal reactions were also mentioned and the authors suggest a different prevalence for the level with ceramics (Silva & Souza, 2001). In Rio Vermelho (Hube, 2005) periosteal reactions and bone infections affect at least 50% of the individuals, children included, 15% of them have osteomielytis.

Prevalence of 60 to 80% of infections in Rio de Janeiro *sambaquis* contrast with 30 to 60% in Santa Catarina. The more severe cases including vault scars, osteitis and osteomielytis are in Rio de Janeiro. Mild cases of fibrous periostitis are in Santa Catarina. In most of the cases benign infection affecting the adults, and there is no relationship with chronology or size of the sites is clear. This first epidemiologic approach is suggestive of a more severe condition in Rio de Janeiro than in Santa Catarina, despite the fact that the bigger sites and populations are in the last region. Climate differences, as well as the natural history of the endemic condition may help to explain the differences. Treponematosis has been suggested as a possible diagnosis.

Under the epidemiological point of view, either different infections or different manifestations of the same infection must also be considered. Detailed comparative studies of the periosteal lesions in different sites will certainly help to propose a paleoepidemiological model. The existence of big groups, low mobility and tropical settlements would favour endemization of a treponemal disease (Powell & Cook, 2005).

MORTALITY

Population size of the *sambaquis* has been proposed on different basis ... and estimated on the amount of shells and other archeological remains. A literature survey in 1973 (Ferraz & Mendonça de Souza, 1973) listed 638 burials in 48 coastal sites, most of them with no children under 5, possibly a taphonomic bias. Although many *sambaquis* have been excavated, part of them are big sites with hundreds of skeletons but were only partially excavated. The substrate usually rich in shells and other calcified debris, is sometimes cemented by the calcium carbonate, making it difficult to recover the bones, especially the fragile and small bones of the youngest ones. Taphonomic factors, among others, possibly explain why the numbers of mortality are so contradictory, and why the children are generally missing, although it is not possible to exclude age related cultural practices in some *sambaquis*.

One of the sites that has been almost completely excavated, providing good data for a paleodemographic study is the *sambaqui* of Piaçaguera, in São Paulo State. Paleodemography of the skeletal series was published by Uchoa, Alvim & Gomes (1987). According to the radiocarbon dates, the site was used for a short period of about 40 years, from 4890 + 110 to 4930 + 110, providing 87 skeletons of adults and children of both sexes, a natural population. 24 are youngs less than 2 years old, 60% of the skeletons are older than 15; 8,7% are older than 55. Life Expectancy (Ex0) was estimated in 21 years, Crude Mortality Rate was 48% and Total Fertility 4,4. Populational Growth was estimated in 0,0015 per year, and the Sex Ratio was well balanced. A bigger mortality of young women between 20-25 years (15,2%) when compared to men of the same age (2,17%) is suggestive of special risks possibly associated to pregnancy and delivery (Hassan, 1981). Once the dates for Piaçaguera are accepted, the short period of use of the cemetery expresses a populational *cohort*. But undoubtedly other big well dated *sambaquis* must still be excavated before we can be conclusive about demographic studies.

The *sambaqui* people is supposed to have lived an affluent life of low mobility, abundant food resources, grouped along the coast in big communities. Perhaps they had some domesticated plants, perhaps mortality has changed during coastal adaptation.

HEALTH AND LIFESTYLE ALONG THE COAST

Neves & Wesolowski (2002) studied eleven *sambaquis* from Babitonga Bay, Santa Catarina State, and used the Health Index proposed by Steckel, Sciulli & Rose (2002). Their results allowed to rank *sambaquis* people as very healthy, reaching between 22,97 and 23,22 points, that is to say about 90% of the maximum health rank to that Index. The results can be a natural consequence of living in a very affluent economy, supported by predictable and productive tropical environments, adding to the fact that living areas would be naturally cleaned by the tidal movements. The economic arguments are apparently correct but the prevalence of infections and stress indicators in published and unpublished data force to think a little more about this interesting problem. It is obvious that a Health Index brings a methodological problem, and the concept of health and disease is subject to discussion as Hubbe (2005) remembers in his thesis. Perhaps the expressive prevalence of infections, the possible endemic condition, the risks of natural infections in salt waters, and the low carbohydrate intake, even in the condition of a hyperproteic diet should be good reasons to be conservative about considering *sambaqui* healthy people.

Paleoepidemiologic data confirm that *sambaqui* groups could have lived in quite different conditions, depending on space and time, suffering from different diseases or biological costs. Living for such a long time and expanding along the coast, they were possibly well adapted, as far as adaptation and health are not necessary togheter. Anemia, infection as well as other problems could be part of their adaptation. Pathocenosis and more detailed biocultural models for sites or groups of sites will be useful to compare different situations for those prehistoric groups and to know more about their health and disease.

FINAL REMARKS

Poor material, a few studies and the lack of curatorial programs are the biggest difficulties to study bioarchaeology of the *sambaqui* people.

Cultural and ecological diversity are the starting points to discuss adaptability and differences in health and adaptative costs among the sambaquis. Diachronic approaches and spatial comparison can help to understand change in lifestyles and in disease. Although big numbers are good, in the case of *sambaqui* sites, small numbers and specific approaches can enlightening specific circumstances before the big numbers are tryied. Some final remarks emerge form this initial data:

1. Caries prevalence is low in most of the *sambaquis* (0 to 2%), lesions are small, inside the fissures and pits, not associated to ceramics. The highest prevalence reach 60%.

2. Tooth wear is intense and can be plane or beveled. In sites with ceramics it is generally anterior. Non mastigatory wear occurs in specific sites.

3. Prevalence of abcesses varies from 12,5 to 70%, there is no correlation with caries or wear/pulp exposure. It may be related to other factors of trauma to periodontal tissues.

4. Dental loss is generally low, except in some cases of possible use of lip ornaments or *tembetás*.

5. Phyotliths and starch of different species insluding in dental calculus are suggestive that high carbohydrate intake could be part of the diet, perhaps domestication, perhaps the exploration of specific natural resources.

6. Anemia as indicated by porotic hyperostosis and *cribra orbitalia* is more prevalent in some sites (86,7%) but rare in others (2 to 5%). The lesions have been active during the first years of life, but they are healed in most of the individuals. There is no sex predominance.

7. Harris' lines are not studied, in Cabeçuda *sambaquis* they confirm a stress period during infancy.

8. The prevalence of linear enamel hypoplasia vary too much, from 64% to 9%. They also confirm physiologic stress between 2-4 years or 4-6 years, depending on the site. This may suggest variation in the weaning period and/or different transitions to the adult nutrition.

9. Infection is represented by periostitis, osteites and osteomielytis affecting 60 to 80% of the individuals in Rio de Janeiro sites, and 30 to 60 % of the individuals in Santa Catarina sites. Endemic patterns, symmetrical lesions, mostly represented by periosteal reactions, and some vault lesions are strongly suggestive that treponemal disease could explain part of the infections at the coastal *sambaquis*.

10. Mortality is not based in consistent numbers, because most of the skeletal series are strongly biased, but the results in Piaçaguera *sambaqui* suggests the mortality was high, among young children (Ex0=21) and young adult women (15,2%).

Under the bioarchaeological point of view, it is possible to confirm the existence of at least two different groups of *sambaquis*. The group of sites in Rio de Janeiro has skeletons with few hypoplasia, low anemia, no torus, paramastigatory wear and high prevalence of infection. The goup of sites in Santa Catarina has skeletons with high prevalence of hypoplasias, anemia, torus, paramastigatory wear and low prevalence of infection. The first data is suggestive that despite time and space, *sambaquis* can be quite different also in biocultural aspects.

Regional difference in paleopathology seems to fit well to differences in archaeological areas. More local pathocenosis should be tested before general models can be proposed, and local differences can be explained by differences in lifestyle.

Bibliography

ALVIM, M.C. de Me, 1978. Caracterização da morfologia craniana das populações pré-históricas do litoral meridional brasileiro (Paraná e Santa Catarina). *Arqui-vos de Anatomia e Antropologia da Universidade Souza Marques 3:* 292-319.

ALVIM, M.C. de Me, VIEIRA, M.I., CHEUICHE, L.M. 1975. Os construtores dos sambaquis de Cabeçuda, SC e de Piaçaguera, SP- estudo morfométrico comparativo. *Arquivos de Anatomia e Antropologia da Universidade Souza Marques I(I):* 395-496.

ALVIM, M.C. de Me, SOARES, M.C. 1984. Incidência de traços não métricos em material de sambaqui do acervo do Museu Nacional da UFRJ. *Revista de Arqueologia 2(1):* 3-12.

ALVIM, M.C. de Me, SOARES, M.C., CUNHA, P.S.P. 1984. Traços não-métricos cranianos e distâncias biológicas em grupos indígenas do Brasil – Botocudos e construtores de sambaquis. *Revista de Pré-História 6:* 107-117.

ALVIM, M.C. de Me, GOMES, J.C. de O. 1989. Análise e interpretação da hiperostose porótica em crânios humanos do Sambaqui de Cabeçuda (SC, Brasil). *Revista de Pré-História 7:* 125-143.

ANGEL, L. 1966. Porotic hyperostosis, anemias, malarias and marshes in the prehistoric eastern Mediteranean. *Science 153:* 760-763.

BUIKSTRA, J.E., COOK, D.C. 1980. Paleopathology: an American account. *Annual Review of Anthropology 9:* 433-476.

COHEN, M.N., ARMELAGOS, G.J. 1984. *Paleopathology at the origins of agriculture*. Accademic Press: New York.

CUNHA, E.S. 1963a. *Sambaquis e outras jazidas arqueológicas. Paleopatologia dentária e outros assuntos*. Editora Científica: Rio de Janeiro.

CUNHA, E.S. 1963b. Afecções alvéolo-dentárias na população do sambaqui de Cabeçuda. *Revista do Museu Paulista (Nova Série) 14:* 523-529.

DANIELS, N.A., MACKINNON, L., BISHOP, R., ALTEPREISE, S., RAY, B., HAMMOND, R.M., THOMPSON, S., WILSON, S., BEAN, N.H., GRIFFIN, P.M., SLUTSKER, L. 2000. *Vibrio parahaemolyticus* infection in the United States, 1973-1998. *Journal of Infectious Diseases 181(5):* 1661-1666.

FERRAZ, S.M., MENDONÇA, de SOUZA, A.A.C. 1973. Paleopatologia do homem dos sambaquis (Notas previas). *Abstracts of the IX International Congress of Anthropological and Ethnological Sciences*. IX ICAES: Washington.

FIGUTI, L. 1993. O homem pré-histórico, o molusco e o sambaqui: considerações sobre a subsistência dos povos sambaquianos. *Revista do Museu de Arqueologia e Etnologia 3:* 67-80.

GASPAR, M.D. 2003. Aspectos da organização social de pescadores-coletores: região compreendida entre a Ilha Grande e o Delta do Paraíba do Sul, Rio de Janeiro. *Pesquisas 59*: 1-163.

GOODMAN, A.H., THOMAS, R.B., SWEDLUND, A.C., ARMELAGOS, G. 1988. Biocultural perspectives on stress in prehistoric, historic and contemporary population research. *Yearbook of Physical Anthropology 31*: 169-202.

GOODMAN, A.H., CAPASSO, L.L. 1992. Recent contribution to the study of enamel developemental defects. *Journal of Paleopathology (Monographic publications #2)*. Associazione Abruzzese: Chiety.

HASSAN, F. 1981. *Demographic archaeology*. Academic Press: New York.

HAUBERT, F, KREVER, M.L.B., PALMA, L.P., SCHMITZ, P.I. 2004. Bocas e dentes: o estudo dos esqueletos escavados por Pe. João Alfredo Rohr, S.J., no litoral de Santa Catarina. *Pesquisas. Série Documentos 9*.

HOOTON, E.A. 1930. *The indians of Pecos Pueblo: a study of their skeletal remains*. Yale University Press: New Haven.

HUBE, M. 2005. *Análise biocultural dos remanescentes ósseos humanos do sambaqui Porto do Rio Vermelho 02 (SC-PRV-02)*. Tese de Doutorado. Instituto de Biociências, Universidade de São Paulo. São Paulo.

KRONE, R. 1908. Exploração ao rio Ribeira do Iguape. Informações ethnográficas do Vale do Rio Ribeira do Iguape. *Boletim da Comissão Geológica de São Paulo 24:* 23-34.

LARSEN, C.S. 1999. *Bioarchaeology. Interpreting behaviour from the skeleton*. Cambridge University Press: Cambridge.

LESSA, A., MEDEIROS, J.C. de. 2001. Reflexões preliminares sobre a questão da violência em populações construtoras de sambaquis: análise dos sítios Cabeçuda e Arapuã. *Revista do Museu Paulista*.

LIMA, T.A. 1991. *Dos mariscos aos peixes: um estudo zooarqueológico da mudança de subsistência na pré-história do Rio de Janeiro*. Tese de Doutorado. Faculdade de Filosofia, Letras e Ciências Humanas, Universidade de São Paulo. São Paulo.

MACHADO, L.M.C., KNEIP, L.M. 1994. Padrões dentários, dieta e subsistência nas populações dos sambaquis de Saquarema. *Revista de Arqueologia 8*.

DeMASI, M.A. 1999. *Pre-historic hunter gatherer mobility on the southern Brazilian coast: Santa Catarina Island*. Tese de Doutorado. Stanford University.

MESSIAS, T.T. 1977. Estudo morfológico da população do Sambaqui do Forte. In: L.M. Kneip *Pescadores e coletores do litoral de Cabo Frio, RJ*. Coleção do Museu Paulista 5 (Série Histórica) Museu Paulista: São Paulo.

NEVES, W.A., UNGER, P., SCARAMUZZA, C.A.M. 1984. Incidência de caries e padrões de subssistência no litoral norte de Santa Catarina, Brasil. *Revista de Pré-história VI:* 371-380.

NEVES, W.A., WESOLOWSKI, V. 2002. Economy, nutrition and disease in the prehistoric coastal Brazil – ma case study from Santa Catarina State. In: R.H. Steckel & J. Rose. *The Backbone of history. Health and nutrition in the Western hemisphere*. Cambridge University Press: Cambridge. 376-400pp.

PEIXOTO, M.V. 1989. *Avaliação radiológica do **torus auditivus** nos grupos formadores de sambaquis do litoral meridional brasileiro: contribuição ao estudo dos trações não métricos em populações pré-históricas do Brasil*. Rio de Janeiro. Dissertação de Mestrado, Departamento de Anatomia Humana, Universidade Federal do Rio de Janeiro. Rio de Janeiro.

POURCHET, M.J., ALVIM, M.C. de Me. 1975. Um estudo de adaptabilidade humana em microregião pesqueira (nota prévia) *Arquivos de Anatomia e Antropologia da Universidade Souza Marques 1(1)*:389-391.

POWELL, M.L., COOK, D.C. 2005. *The myth of syphilis. The natural history of treponematosis in North América*. University Press of Florida: Gainesville.

PROUS, A. 1992. *Arqueologia brasileira*. Editora da Universidade de Brasília: Brasília.

REINHARD, K.J., SOUZA, S.M.F.M. de, RODRIGUES, C., KIMMERLE, E., DORSEYVINTON, S. 2001. Dental calculus a new perspective on diet and disease. In: *Human Remains, Conservation and Retrieval and Analysis*. British Archaeological Research Council: London.

RODRIGUES, C., IMÁZIO, M., SILVA, E.C. da, e SILVA, L.G.L. da. 2001. Os remanescentes esqueletais do sambaqui do Moa – Saquarema, RJ – recuperados na escavação de salvamento de 1998: dados preliminares. In: Arno Alvarez Kern & Klaus Hilbert, *Arqueologia do Brasil Meridional*. PUCRS: Porto Alegre (CD ROM).

RODRIGUES-CARVALHO, C., SOUZA, S.M.F.M. de. 1998. Uso de adornos labiais pelos construtores do sambaqui de Cabeçuda, Santa Catarina: uma hipótese baseada no perfil dento-patológico. *Revista de Arqueologia Brasileira* 11: 43-55.

RODRIGUES-CARVALHO, C. 2004. *Marcadores de estresse ocupacional em populações sambaquieiras do litoral fluminense*. Tese de Doutorado. Escola Nacional de Saúde Pública/Fiocruz. Rio de Janeiro.

SILVA, S.F.S.M. da. 2005. *Arqueologia das práticas mortuárias em sítios pré-históricos do Estado de São Paulo*. Tese de Doutorado. Museu de Arqueologia e Etnologia, Universidade de São Paulo. São Paulo.

SCHEEL, R., GASPAR, M.D., YBERT, J.P. 1996. Antracologia: uma nova fonte de informação para a arqueologia brasileira. *Revista do Museu de Arqueologia e Etnologia 6:* 3-9.

SILVA, E.C.da, SOUZA, S.M.F.M. de. 2001. Paleopatologia do Forte Marechal Luz, SC, Brasil:

reabertura de arquivo. *Arqueologia do Brasil Meridional (*CD ROM*).* Pontifícia Universidade Católica do Rio Grande do Sul: Porto Alegre.

SOUZA, S.M.F.M. de. 1992. Um caso pré-histórico de osteomielite: treponematose ou infecção generalizada. In Maria da Conceição Beltrão, *Arqueologia do Estado do Rio de Janeiro*. Arquivo Nacional: Rio de Janeiro. 79-94pp.

SOUZA, S.M.F.M. de. 1995. *Estresse, doença e adaptabilidade: estudo comparativo de dois grupos pré-históricos em perspectiva biocultural.* Tese de Doutorado. Escola Nacional de Saúde Pública/ Fiocruz: Rio de Janeiro.

SOUZA, S.M.F.M. de. 1999. Anemia e adaptabilidade em um grupo costeiro pré-histórico: uma hipótese patocenótica. In Maria Cristina Tenório, *Pré-história de Terra Brasilis*. EDUFERJ: Rio de Janeiro. Pp171-188.

SOUZA, S.M.F.M. de, CARVALHO, D.M., LESSA, A. 2003. Paleoepidemiology: is there a case to answer? *Me-mórias do Instituto Oswaldo Cruz 98 (Supl.):* 21-27.

SOUZA, S.M.F.M. de, WESOLOWSKI, V, RODRIGU-ES-CARVALHO, C., LESSA, A. 2006. Bioarchaeology of the Brazilian shellmound people: helping to understand differences in lifestyle. *Abstracts of the 16th European Meeting of the Paleopathology Association*. PPA: Santorini.

STECKEL, R.H., SCIULLI, P.W., ROSE, J.C. 2002. A Health index from skeletal remains. In: Richard H. Steckel & Jerome Rose. *The backbone of history. Health and nutrition in the Western hemisphere*. Cambridge University Press: Cambridge. 61-93.

STORTO, C., EGERS, S., LAHR, M.M. 1999. Estudos preliminares das paleopatologias da população do sambaqui de Jaboticabeiras II, Jaguaruna, SC. *Revista do Museu de Arqueologia e Etnologia 9:*61-74.

STORTO, C., EGGERS, S., LAHR, M.M. 2001. Os construtores do sambaqui de Jaboticabeira II. In: Arno Alvarez Kern & Klaus Hilbert, *Arqueologia do Brasil Meridional*. PUCRS: Porto Alegre. (CD ROM).

STUART-MACADAM, P., KENT, S. 1992. *Diet, demography and diseases. Changing perspectives in anemia*. Aldine: New York.

UCHOA, D.P., ALVIM, M.C. de Me, GOMES, J.C. de O. 1987. Demografia esqueletal dos "construtores" do sambaqui de Piaçaguera, São Paulo, Brasil. *Anais da IV Reunião Científica da Sociedade de Arqueologia Brasileira*. Sociedade de Arqueologia Brasileira: São Paulo.

WESOLOWSKI, V. 2000. *A prática da horticultura entre os construtores dos sambaquis e acampamentos litorâneos da região da Baía de São Francisco, SC. Uma abordagem bioantropológica*. Dissertação de Mestrado. Faculdade de Filosofia Letras e Ciência Humans, Universidade de São Paulo. São Paulo.

WESOLOWSKI, V.; REINHARD, K.J.; SOUZA, S.M.F.M. de; CECCANTINNI, G. 2007. Grânulos de amido e fitólitos em cálculos dentários humanos. Contribuições ao estudo do modo de vida e subsistencia dos grupos sambaquianos do litoral sul do Brasil. *Revista do Museu de Aquilegia e Etnologia (São Paulo)* 17: 191-210.

THE CONTRIBUTION OF CRANIAL MORPHOLOGY OF HUMAN SKELETAL REMAINS TO THE UNDERSTANDING OF THE BIOLOGICAL AFFINITIES BETWEEN COASTAL AND RIVERINE SHELLMOUNDS IN SOUTHERN BRAZIL

Maria Mercedes M. OKUMURA*, Walter A. NEVES

Laboratório de Estudos Evolutivos Humanos, Depto. de Genética e Biologia Evolutiva, Instituto de Biociências, Universidade de São Paulo, CP 11461, 05422-970, São Paulo, SP, Brazil

* New address: Leverhulme Centre for Human Evolutionary Studies, The Henry Wellcome Building, University of Cambridge, Fitzwilliam Street, Cambridge CB2 1QH, UK. mmo23@cam.ac.uk

***Abstract**: Brazilian shellmounds are sites found along the entire coast and in some adjacent riverine regions in this country. Here, we use craniometrical data in order to find out if there is a biological connection between these two populations. We analyzed 9 males and 3 females from 5 riverine shellmounds from São Paulo and Paraná State (Moraes, Itaoca 1, Estreito 1, Pavão XVI and Pavão III), dated to between 5.900 and 1.200 yBP. Principal Component Analyses (corrected for size) pointed to a certain degree of similarities between the coastal and riverine populations, ate least at the macro-zone of the Ribeira do Iguape valley.*

***Key Words**: craniometry, Ribeira de Iguape valley, Principal Component Analysis*

***Resumo**: Sambaquis são sítios arqueológicos encontrados ao longo de quase toda a costa brasileira e ao longo de alguns rios. Neste trabalho, analisamos 12 esqueletos (9 homens e 3 mulheres) exumados em cinco sambaquis fluviais do vale do Ribeira de Iguape datados entre 5.900 e 1.200 anos AP. Análises de Componentes Principais (com o fator tamanho corrigido) apontam para um certo grau de similaridade entre grupos costeiros e fluviais, ao menos na macro-zona do vale do Ribeira de Iguape.*

INTRODUCTION

The degree of contact between inland and coast in the southern Brazilian region during the prehistory is still a polemic issue. Although some authors proposed an independent development for the coast and for the inland (Altenfelder Silva, 1967; Laming-Emperaire, 1975), most scholars agree with the idea of contact between the coast and the interior (Barreto, 1988; Schimtz *et al.*, 1993; Schmitz, 1996; Willey, 1971). In Southern Brazil, the presence of typical material culture and food remains of coastal origin in inland sites (Barreto, 1988; Figuti *et al.*, 2004; Tiburtius *et al.*, 1950-1951), as well as the presence of the same ceramic traditions in late times, also contribute to strength the idea of a relatively significant contact between inland and coast (Araujo, 2001; Batista da Silva *et al.*, 1990; Beck, 1971; Prous, 1979; Reis, 1980; Rohr, 1971).

The communication between coast and inland in southern Brazil is not an easy task due to the "Serra do Mar". However, the Ribeira de Iguape valley, on the southeast of São Paulo State is one of the few places where the communication could easily happen. In this study, we focus on the riverine shellmounds found in the Ribeira de Iguape valley to test if they could be related to the neighbour coastal shellmounds.

The name "riverine shellmounds" is due to the massive presence of terrestrial mollusc shells, like *Megalobulimus* sp. and *Strophocheilus* sp., in these sites, and the proximity of a river. They are shallow (50 cm to one meter of height), composed by sediment with mollusk shells sparced or concentrated in pits (Collet & Prous, 1977). The lithic and the bone industries of the riverine shellmounds are very similar to those found in the coast. Marine faunal remains are also found in several archaeological sites of Ribeira de Iguape valley (Barreto, 1988; Figuti *et al.*, 2004), attesting to interactions between the coast and the interior.

The idea that the coastal shellmounds are the product of one or more migrations from the interior is very popular among Brazilian archaeologists, however, there are no archaeological evidences of how and when this process begun, and which routes would had been used. Barreto (1988) proposed an inverse model, in which groups inhabiting the coast would have move to the interior, keeping the habit of building shellmounds. For Collet and Prous (1977) the riverine shellmounds of Ribeira de Iguape valley would be temporary settlements established by the coastal groups.

In this study, we present the results of a morphological comparative analysis involving human skeletons recovered from several riverine shellmounds from Ribeira de Iguape valley, and from the adjacent shoreline.

MATERIAL AND METHODS

Twelve skeletons recovered from several riverine shellmounds were tested in this work. Dates available for each site are shown in Table 7.1. The riverine shellmound skeletons were compared to some Brazilian skeletal series from the interior and the coast in order to access the degree of biological affinities. This comparison was conducted through Principal Component Analysis (PCA)

Tab. 7.1. Non calibrated C^{14} dates from riverine shellmounds used in this study. All dates obtained from human bone material, except Itaoca 1 (after Figuti *et al.*, 2004)

Archaeological site	Structure	Depth (cm)	Noncalibrated age	Reference
Moraes	Sep 25	25	4511 ± 32	KIA* 20844
Moraes	Sep 35	35	5420 ± 30	KIA 20843
Moraes	Sep 5	100	4985 ± 35	KIA 15562
Moraes	Sep 13	130	5895 ± 45	KIA 15561
Itaoca 1	S1	20-30	1460 ± 60	Beta** 178128
Estreito 1	Sep 1	25	3655 ± 26	KIA 20845
Estreito 1	Sep 6	130	4124 ± 27	KIA 20846
Pavão XVI	Sep 1	30	1571 ± 24	KIA 20842
Pavão III	Sep 1	20	1219 ± 24	KIA 20840

* Kiel AMS, Germany
** Beta Analytic, USA

Tab. 7.2. Samples used (males)

Series	Series composition	Geographic location	Oldest date	N for each site	N total
Moraes	Idem	Ribeira de Iguape valley, SP	5895 ± 45	-	4
Riverine composite	Itaoca 1	Ribeira de Iguape valley, SP	1460 ± 60	1	5
	Estreito 1	Ribeira de Iguape valley, SP	4124 ± 27	1	
	Pavão III	Ribeira de Iguape valley, SP	1219 ± 24	2	
	Pavão XVI	Ribeira de Iguape valley, SP	1571 ± 24	1	
Sumidouro	Idem	Lagoa Santa, MG	> 8000	-	17
Umbu	Idem	RS	-	-	3
Estirão Comprido	Idem	Prudentópolis, PR	-	-	4
Paraná Interior	Abrigo do Pontão	Sengés, PR	-	2	3
	Morro dos Anjos	Adrianópolis, PR	-	1	
Guaraguaçu	Idem	Paranaguá, PR	4200 ± 200	-	20
Paraná Coast	Matinhos	Matinhos, PR	2750 ± 250	10	25
	Centenário	Paranaguá, PR	-	3	
	Macedo	Paranaguá, PR	3677 ± 61	3	
	Ilha dos Ratos	Guaratuba, PR	1540 ± 150	2	
	Araújo II	Guaratuba, PR	-	1	
	Guaratuba (region)	Guaratuba, PR	-	1	
	Rio Laranjeiras	Guaratuba, PR	-	1	
	Toral 51	Paranaguá, PR	-	1	
	Ilha Rasa / da Gamela (region)	Guaraqueçaba, PR	-	1	
	Ponta do Goulart (region)	Antonina, PR	-	2	
SPaulo Northern coast	Tenório	Ubatuba, SP	1875 ± 90	3	5
	Mar Virado	Mar Virado Island, SP	2640 ± 70	2	
SPaulo Central coast	Piaçaguera	Cubatão, SP	4930 ± 110	17	42
	Ilha de Santo Amaro (region)	Santo Amaro Island, SP	-	12	
	Buracão	Santo Amaro Island, SP	2050 ± 100	6	
	Maratuá	Santo Amaro Island, SP	3865 ± 95	4	
	Casqueirinha	Santos, SP	-	1	
	Santos (region)	Santos, SP	-	1	
	Cosipa 4	Santos, SP	2590 ± 80	1	

Series	Series composition	Geographic location	Oldest date	N for each site	N total
SPaulo Southern coast	Boguassu	Ilha Comprida, SP	-	4	12
	Iguape (region)	Iguape, SP	-	4	
	Brocoanha II	Baixada Cananéia-Iguape, SP	-	2	
	Boa Vista	Ilha Comprida, SP	-	1	
	Cananéia (region)	Cananéia, SP	-	1	
Total					140

Tab. 7.3. Samples used (females)

Séries	Series composition	Geographic location	Oldest date	N for each site	N total
Moraes	Idem	Ribeira de Iguape valley, SP	5895 ± 45	-	3
Sumidouro	Idem	Lagoa Santa, MG	> 8,000	-	13
Estirão Comprido	Idem	Prudentópolis, PR	-	-	3
Paraná Interior	Abrigo do Pontão	Sengés, PR	-	4	7
	Morro dos Anjos	Adrianópolis, PR	-	3	
Guaraguaçu	Idem	Paranaguá, PR	4200 ± 200	-	19
Paraná Coast	Matinhos	Matinhos, PR	2750 ± 250	11	19
	Ilha dos Ratos	Guaratuba, PR	1540 ± 150	3	
	Rio da Praia	Guaratuba, PR	-	1	
	Boguassu	Guaratuba, PR	-	1	
	Macedo	Paranaguá, PR	3677 ± 61	1	
	Barra do Veríssimo (region)	Paranaguá, PR	-	1	
	Saquarema	Morretes, PR	4438 ± 89	1	
SPaulo Northern coast	Tenório	Ubatuba, SP	1875 ± 90	-	7
SPaulo Central coast	Buracão	S. Amaro Island, SP	2050 ± 100	5	15
	Maratuá	S. Amaro Island, SP	3865 ± 95	4	
	Santo Amaro Island (region)	S. Amaro Island, SP	-	4	
	Piaçaguera	Cubatão, SP	4930 ± 110	2	
SPaulo Southern coast	Boguassu	Ilha Comprida, SP	-	5	11
	Brocoanha II	Baixada Cananéia-Iguape, SP	-	3	
	Boa Vista	Ilha Comprida, SP	-	2	
	Iguape (region)	Iguape, SP	-	1	
Total		-	-		97

based on 47 cranial measurements (male sample) and 25 measurements (female sample), all described by Howells (1973; 1989). Due to the poor preservation of many individuals, specimens from similar sites had to be grouped. For the same reason, means were used instead individual values. The number of individuals included is shown in Tables 7.2 (male data) and 7.3 (female data).

The PCA were undertaken based only on shape alone. Q standardization (Darroch & Mosiman, 1985) was used to eliminate size influence. The biological affinities of riverine groups were explored through cluster analysis (Ward's method) applied to Euclidian distances (Ward, 1963). Distances were calculated from the scores of the first six PC (male sample) and from four PC (female sample), in which the relevant information was concentrated (Johnson & Wichern, 2002). Cluster analysis was used instead of bi-dimensional graphics because only near 50% of the original information were retained by the first two PCs in both analysis (male and female sample).

RESULTS

The first six PC extracted for the male sample explain about 88% of the original variation. Clearly, two main groups can be defined in Figure 7.1: a group formed by Sumidouro and Umbu and another one, aggregating the

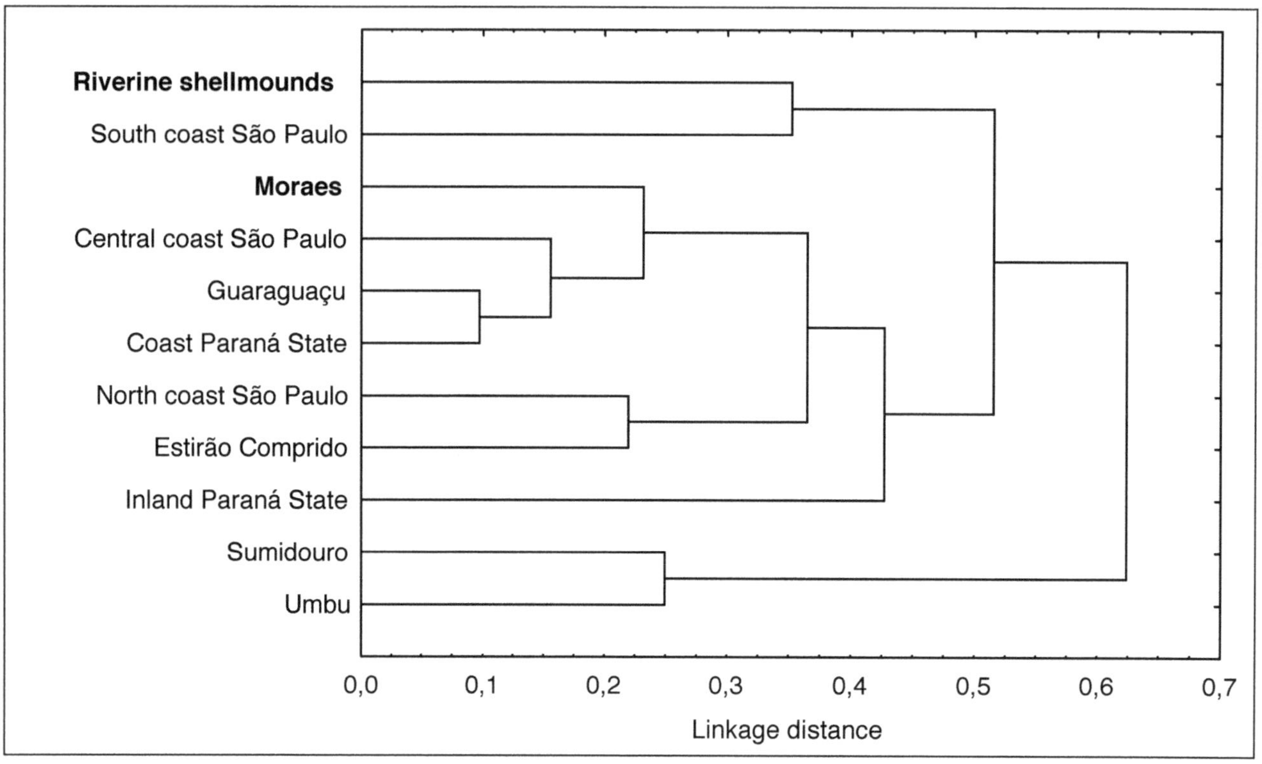

Fig. 7.1. Dendrogram based on Ward's method (male)

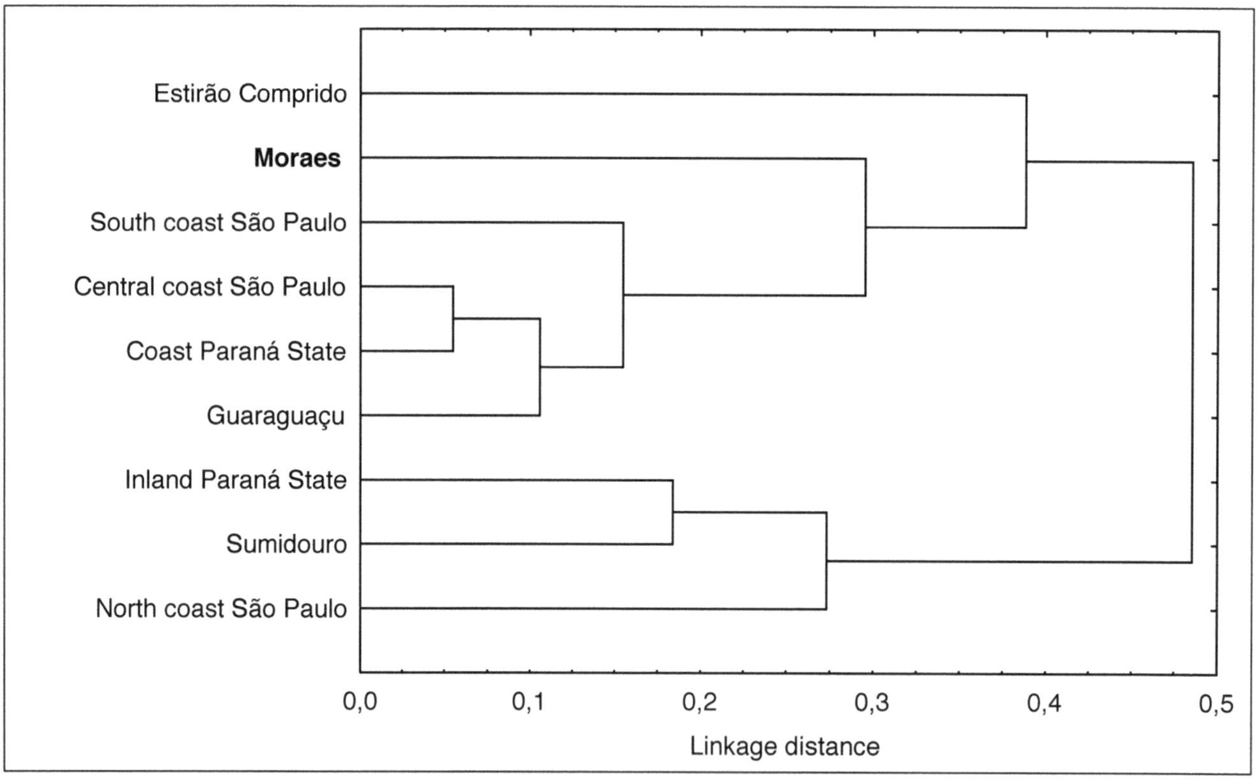

Fig. 7.2. Dendrogram based on Ward's method (female)

rest of the series, including the ones coming from Ribeira de Iguape valley. Within this second group, Moraes is clearly associated to Central coast of São Paulo, coast of Paraná, and Guaraguaçu (a coastal shellmound also from Paraná), while riverine composite is associated to Southern coast of São Paulo.

As to the females the first four PCs explain about 87% of the original information. Figure 7.2 presents the cluster based on the first four PC scores. Again, two distinct groups are distinguished: one formed by North coast of São Paulo, Inland Paraná and Sumidouro; another one formed by the remaining groups. Moraes, the only riverine sample available for this analysis, presents more affinity to south coast of São Paulo, Guaraguaçu, coast of Paraná, and central coast of São Paulo.

DISCUSSION

Our results support a biological relationship between the riverine shellmounds from Ribeira de Iguape and the coastal shellmounds from São Paulo (South and Central coast), and Paraná.

The existence of biological affinities between the coast and the inland of São Paulo supports the hypothesis of an important gene flow between the two regions. This agrees with the similarities in material culture, and, obviously, with the construction of an elevated platform to be used as a residential and funerary space. As already mentioned, Ribeira de Iguape valley is one of the few places in Southern Brazil where the geographical transition between shore and inland could have facilitated and stimulated contacts between human groups inhabiting both regions. It remains to be solved, however, if the original dispersion was from the inland to the seashore, or vice-versa.

As the riverine skeletons analyzed in this article are dated to from 6.000 to 1.200 years BP, it is possible to say that at least from the Middle Holocene on people inhabiting Ribeira de Iguape valley exhibited a classic mongoloid cranial shape. As in other several regions of South America, they apparently replaced their paleoamerican predecessors (non-mongoloid morphology) who were present in the region in the Early Holocene (Neves *et al.*, 2005). Unfortunately, the lack of human skeletons from Ribeira do Iguape dated to from 8.800 to 6.000 years BP prevents a precise determination of when the replacement occurred.

Acknowledgements

We thank the staff of the following institutions for allowing us to examine human skeletons in their care:

- Laboratório de Antropologia Biológica (IB-USP),
- Museu de Arqueologia e Etnologia (USP-SP),
- Museu Nacional (UFRJ-RJ),
- Museu Paranaense (PR),
- Centro de Estudos e Pesquisas Arqueológicas (UFPR-PR),
- Museu de Arqueologia e Etnologia (UFPR-PR),
- Museu Arqueológico de Sambaqui de Joinville (SC),
- Museu Arqueológico do Rio Grande do Sul (MARSUL-Rio Grande do Sul),
- Zoological Museum (University of Copenhagem).

This research was supported by FAPESP (Grants 02/13441-0 and 04/01321-6) and CNPq (305918/00-85).

References

ALTENFELDER SILVA, F. (1967) – Informes preliminares sobre a arqueologia de Rio Claro. PRONAPA 1, Resultados preliminares do primeiro ano (1965/66). *Publicações Avulsas*. Museu Paraense Emílio Goeldi. 6.

ARAUJO, A.G.M. (2001) – *Teoria e método em arqueologia regional: um estudo de caso no alto Paranapanema, Estado de São Paulo*. São Paulo, PhD dissertation, Universidade de São Paulo.

BARRETO, C.N.G.B. (1988) – *A ocupação pré-colonial do Vale do Ribeira de Iguape, SP: os sítios concheiros do médio curso*. São Paulo, PhD dissertation, Universidade de São Paulo.

BATISTA DA SILVA, S.; SCHMITZ, P.I.; ROGGE, J.H.; DE MASI, M.A.N.; JACOBUS, A.L. (1990) – Escavações arqueológicas do Pe. João Alfredo Rohr, s.j. – O sítio arqueológico da praia da Tapera: um assentamento Itararé e Tupiguarani. *Pesquisas (Antropologia)*. 45, p. 1-210.

BECK, A. (1971) – Grupos cerâmicos do litoral de Santa Catarina. Fase Rio Lessa e Fase Enseada. *Anais do Museu de Antropologia UFSC*. 4, p. 25-29.

COLLET, G.; PROUS, A. (1977) – Primeiro informe sobre os sambaquis fluviais da região de Itaoca (SP). *Arquivos do Museu de História Natural*. Belo Horizonte. 2, p. 31-35.

DARROCH, J.N.; MOSIMANN, J.E. (1985) – Canonical and principal components of shape. *Biometrika*. 72: 2, p. 241-252.

FIGUTI, L.; EGGERS, S.; MENDONÇA, C.A.; PORSANI, J.L.; ROCHA, E.B.; DE BLASIS, P.A.D.; BISSA, W.M. – *Investigações Arqueológicas e Geofísicas dos sambaquis fluviais do vale do Ribeira de Iguape, Estado de São Paulo* [Manuscrito]. 2004. Acessível na Biblioteca do Museu de Arqueologia e Etnologia, São Paulo, Brasil.

HOWELLS, W.W. (1973) – *Cranial variation in man*. Papers from the Peabody Museum of Archaeology and Ethnology 67. Cambridge, Massachusetts: Peabody Museum of Archaeology and Ethnology.

HOWELLS, W.W. (1989) – *Skull shapes and the map: craniometric analyses in the dispersion of modern Homo*. Papers of the Peabody Museum of Archaeology and Ethnology 79. Cambridge, Massachusetts: Peabody Museum of Archaeology and Ethnology.

JOHNSON, R.A.; WICHERN, D.W. (2002) *Applied multivariate statistical analysis*. Upper Saddle River, N.J.: Prentice Hall.

LAMING-EMPERAIRE, A. (1975) – Problèmes de préhistoire brésilienne. *Annales Économies, Sociétés et Civilizations*. 5.

NEVES, W.A.; HUBBE, M.; OKUMURA, M.M.M.; GONZÁLEZ-JOSÉ, R.; FIGUTI, L.; EGGERS, S.; DE BLASIS, P.A.D. (2005) – A new early Holocene human skeleton from Brazil: implications for the settlement of the New World. *Journal of Human Evolution*. 48. p. 403-414.

PROUS, A. (1979) – Premiére information sur les maisons souterraines de l'Etat de São Paulo. *Revista de Pré-História*. São Paulo. 1:1. p. 127-145.

REIS, M.J. (1980) – *A problemática das estruturas subterrâneas no Planalto Catarinense*, São Paulo. Master thesis, Universidade de São Paulo.

ROHR, J.A. (1971) – Os sítios arqueológicos do Planalto Catarinense, Brasil. *Pesquisas (Antropologia)*. 24. p. 1-56.

SCHMITZ, P.I. (1996) – Visão de conjunto dos sítios de Tapera, Armação do Sul, Laranjeiras I e II, Pântano do Sul e Cabeçudas – Escavações arqueológicas do Pe. João Alfredo Rohr, S.J. *Pesquisas (Antropologia)*. 53. p. 183-190.

SCHMITZ, P.I.; VERARDI, I.; DE MASI, M.A.N.; ROGGE, J.H.; JACOBUS, A.L. (1993) – O sítio da Praia das Laranjeiras II. Uma aldeia da tradição ceramista Itararé – Escavações arqueológicas do Pe. João Alfredo Rohr, S.J. *Pesquisas (Antropologia)*. 49. p. 1-181.

TIBURTIUS, G.; BIGARELLA, I.K.; BIGARELLA, J.J. (1950/1951) – Nota prévia sobre a jazida paleontográfica de Itacoara (Joinville, Estado de Santa Catarina). *Arquivos de Biologia e Tecnologia*. 5/6. p. 315-340.

WARD, J.H. (1963) – Hierarchical grouping to optimize an objective function. *Journal of the American Statistical Association*. 58. p. 236-244.

WILLEY, G.R. (1971) – *An introduction to American archaeology: South America* 2. Englewood Cliffs, N.J.: Prentice Hall.

SAMBAQUIS THE BRAZILIAN SHELL MOUNDS: WHAT IS THAT ALL ABOUT?

Sheila Mendonça de SOUZA
Departamento de Endemias, Escola Nacional de Saúde Pública Sérgio Arouca, Fundação Oswaldo Cruz, Brasil, sferraz@ensp.fiocruz.br

Claudia Rodrigues CARVALHO
Setor de Antropologia Biológica, Departamento de Antropologia, Museu Nacional, Universidade Federal do Rio de Janeiro, Brasil, claudia@mn.ufrj.br

***Abstract**: Brazilian sambaquis, as other shell mounds all around the world, are prehistoric sites that represent extraordinary strategies of coastal adaptation. They are intentionally build and preserve cultural as well as biological remains. Thousands of burials offer precious material for bioarchaeological investigation. Differences in cultural and environmental conditions along 6.000 years, as well as a peculiar geographical distribution along the southeastern coast of Brazil, make these sites interesting subjects to investigate adaptability and health. Their disapearance was about 1.300 years before present. Their possible contact with ceramists and other people are still to be explained.*
***Key words**: Sambaqui, shell mound, Brazil, coastal adaptation, funerary monuments*

***Resumé**: Les sambaquis brésiliens, comme d'autres amas coquilles du monde, sont parmi les groupes prehistoriques du Brèsil qui signalent le développement d'une estrategie extraordinaire d'adaptation au litoral. Sa construction fut intentionnel et c'est possible y rencontrer des vestiges culturels ainsi que biologiques. Les milliers de sépultures preservées aux sambaquis sont une source assez importante pour la recherche bio-archéologique. La distribution geographique peculiaire, au long du litoral sud et sud-est, ainsi que les changements culturels et d'environnement pendant 6.000 d'annés, faisent les sambaquis très interresants pour les investigations sur la santé et l'adaptabilité. Sa disparition à l'envion 1300 ans BP et la possibilité du contact aux groupes ceramistes sont encore sujets qui restent a expliquer.*
***Mots clés**: Sambaqui, amas de coquilles, Brezil, adptation au litoral, monument funeraire*

The symposium BIOARCHAEOLOGY AT THE MIDST OF SHELLS was proposed to discuss similarities and differences between the shell mounds in Portugal and Brazil. To help the participants discussing the proposed theme the organizers decided to add brief comments to the oral presentation presenting the state of art on the archaeology of the marine and riverside shell mounds. This paper reports the archaeology of the Brazilian marine *sambaquis,* emphasizing the interest in recovering and analyzing more bioarchaeological data about the people who built those mounds.

Sambaqui is the local name for Brazilian shell mounds, a very common type of prehistoric site. Similar to other ones that are found all around the world, they are also burial places. Preserving human skeletons, the *sambaquis* are extraordinary windows that let us know more about lifestyle and health in the past. In Brazil the *sambaquis* does exist especially along the southeastern coast, as an expression of a new economic strategy of intense exploitation of the environments close to the estuaries, lagoons, mangroves and sea. The biggest marine *sambaquis* are found in Santa Catarina State, but others of smaller sizes are also found in Rio de Janeiro, São Paulo, Paraná and Espírito Santo States, where more than one thousand of them have already been registered. Most of them have been destroyed before any research. About 5% of them were minimally excavated and less than 1% provided human remains for scientific studies.

These sites are composed by layers of mollusk shells, fish bones and other faunal remains charcoal, polished as well as flaked lithic instruments as axes, grinding stones, arrow points among others. Archeological structures such as hearths and burials with different types of funerary offerings are also found (Prous, 1992; Figuti, 1993; Gaspar, 2000; Gaspar *et al.*, 2008). Although excavated for more than one hundred years, there is no consensus among scholars about them and classifications are based in different criteria such as morphology, geographical location, shell composition, etc. Zooliths, beautiful polished stone sculptures representing animals are found in some of them (Prous, 1992). Ceramics is present in a small number of sites, grouped in the same area. The interpretation of sambaqui has changed through time. First they were viewed as midden left over by small mollusk eating nomadic bands. Today they are seen as monumental constructions intentionally built as landmarks by sedentary fisher groups (DeBlasis XXX, Figuti, XXX, DeMasi, XXX).

Despite temporal and regional variation, there are unifying factors among the sambaquis that allow to try to interpret them as a "cultural unit", as proposed by Gaspar (1994/1995). Eating, feasting, working, burying the dead and daily activities were carried out at the same sopot. Differences among the sites are not necessary because of cultural differences but they may express adaptability along the coast, changing economy, feeding habits and

Fig. 8.1. Sambaqui Cubatão I, at Joinville, Santa Catarina State. Field season of july 2008. The excavation is part of a biocultural project of the authors (photo: Dione Bandeira)

even the use of certain artifacts, as strongly suggested by the variability even inside the same site.

Today the *sambaquis* are recognized as the expression of intentional accumulation of shells and soil to build platforms. In Brazil these territorial marks can reach 30m high and 400m length (Fig. 8.1). Such a building effort certainly required a big amount of energy, a lot of social organization. Big groups and long periods of continuous occupation occurred in some sites. There are hundreds of burials in the biggest *sambaquis*. Most of the sambaquis are grouped in coastal areas signaling to the areas rich in natural resources and fresh water supply, but also to some king of social articulation. Archaeological remains proved that fishing, collecting shellfish/plants and chasing terrestrial and marine animals were the main economic activities. Low mobility, abundance of resources, bigger groups, some social complexity and possible social interaction between sites is supposed to have occurred (Lima, 1999-2000). Ideological, aesthetic and religious development is inferred from the burials.

Coastal environment provides good and abundant proteins for food. The Atlantic forest and the *restingas* provide fruits and other starch sources. The existence of predictable natural resources along the coast was a positive factor to explain the success of *sambaqui* lifestyle. Sharing simple and efficient artifacts made of mammal and shark teeth, fish and mammal bones, shells, simple flaked rock crystals and pebble stones, the well succeeded *sambaquis* expanded in time and space. Amazing polished stone artifacts (Fig. 8.2) in the graves point to cultural variability along the coast. As of now, about 1.600 skeletons were recovered in 60 different Brazilian sites. One of the biggest questions for bioarchaeologists today is if they are able to elucidate the unity and diversity of that people. For many decades questions about the biological characteristics of the *sambaqui* people have been asked (Alvim, 1978). New answers to ancient questions come from the studies on paleopathology, paleodiet, paleodemography.

Geographical distribution of the *sambaquis* and their temporal persistence (8.000 to 1.300 BP). Brazilian sites distinguish from sites in other parts of the world, this richness points to a well succeeded strategy in tropical environments. At the same time it helps to explain variability in culture and biological cost of each lifestyles.

Fig. 8.2. Sculptures made of polished stones representing coastal fauna are characteristics of Brazilian shellmounds

These characteristics are guidelines to discuss paleoepidemiological models for *sambaquis*. Their adaptability was proved by the ability to explore different ecotones along the productive and predictable coastal waters, resulting in the intense exploration of the mangroves, beaches, shallow water lagoons, rocky coastal reefs, sandy plains and estuarine areas as fishing-hunting-collecting fields.

Holocene *sambaqui* lifestyle had the climax between 5.000 and 3.000 y.b.p. The consequences of the contact with other prehistoric people is not clear, in special the ceramists that arrived at the coast 800 to 1.300 years before present (Gaspar, 1994-1995). When the villages of the Itararé tradition and the Tupi and Guarani finally dominated the coast, *sambaqui* people was not there any more. For some authors, lifestyle based on the water born resources was simply adapted to include domestication of plants like manioc, maize and yams. The systematic use of seasonal natural carbohydrate provided by seasonal sources such as Brazilian pine could happen along with the water protein intake in coastal environments. New data on paleonutrition may help to elucidate this hypothesis in the next future (Weslowski *et al.*, 20007). A great number of coastal sites are not characteristic *sambaquis* but share common artifacts and economic characteristics, suggesting that changes and diversity may have just been adaptative. Village settlements with or without ceramics can be found especially along the last millennia in the areas where *sambaquis* were dominant 4.000 or more years before. Comparative studies of groups living in true *sambaquis* and other sites will possibly help to elucidate their possible biological relationships.

Concerning our present research, some questions are in the base of the bioarchaeology of the *sambaquis:* Were all the *sambaqui* groups similar? What was the biologic impact o their cultural differences? What was the impact of their lifestyles in health? Were they really well adapted as supposed? Which were their biological relationships to other prehistoric people?

Referências bibliográficas

ALVIM, M.C. de Me. 1978. Caracterização da morfologia craniana de populações pré-histõricas ddo litoral meridional brasileiro (Paraná e Santa Catarina). *Arquivos de Anatomia e Antropologia do Instituto Souza Marques (Rio de Janeiro)* III: 293-318.

De MASI, M.A.N. 1999. Prehistoric hunter gatherer mobility on the southern brazilian coast: Santa Catarina Island. Tese de Doutorado: Stanford University.

FIGUTI, L. 1993. O homem pré-histórico, o molusco e o sambaqui: considerações sobre a subsistência dos povos sambaquianos. *Revista do Museu de Arqueologia e Etnologia (São Paulo)* 3: 67-80.

GASPAR, M.D. 1994-1995. Espaços, ritos funerários e identidade pré-histórica. *Revista do Museu de Arqueologia e Etnologia (São Paulo)* 8(2):221-237.

GASPAR, M.D.; De BLASIS, P.; FISH, S.K.; FISH, P.R. 2008. Sambaqui (Shell mound) societies of coastal Brazil. In: Helaine Silverman & William H. Isabelli (Eds.) Handbook of South American Archaeology. Springer: p. 319-338.

LIMA, T.A. 1999-2000. Em Busca dos Frutos do mar: os Pescadores-coletores do Litoral Centro-sul do Brasil. Revista USP, 44:270-332.

PROUS, A. 1992. Arqueologia brasileira. Editora da Universidade de Brasília: Brasília.

WESOLOWSKI, V.; SOUZA, S.M.F. M. de; REINHARD, K.J.; CECCANTINNI, G. 2007. Grânulos de amido e fitólitos em cálculos dentários humanos> contribuições ao estudo do modo de vida e subsistência dos grupos sambaquianos do litoral sul do Brasil. *Revista do Museu de Arqueologia e Etnologia (São Paulo)* 17: 191-210.

RIVERINE VERSUS COASTAL SHELLMOUNDS IN BRAZIL

Sabine EGGERS

Laboratório de Antropologia Biológica, Depto Genética e Biologia Evolutiva, Inst Biociências, Universidade de São Paulo, Rua do Matão 277, 05508-900 São Paulo, Brazil, e-mail: saeggers@usp.br

Abstract: *In this paper we present a general overview on individuals recently exhumed from riverine shellmounds (Vale do Ribeira, São Paulo State, Brazil) in comparison with reports on skeletal collections of coastal shellmound builders housed in different institutions. We begin with a brief contextualization describing similarities and differences regarding location, date, size, subsistence and cultural features. Finally we adress palaeopathological markers. We conclude that health and way of life of the people of the fluvial sambaqui Moraes lies in the range of distribution of paleopathologies observed among coastal sambaqui groups, except for their high neck caries frequency. These results suggest mainly differences in subsistence strategies. These differences are explored in Boyadjian* et al. *(this issue) and shed new light on the way of life of the people who built the riverine sambaquis in the vale do Ribeira.*

Key words: *fluvial shellmounds, sambaquis, site implantation, cultural affinities, subsistence, palaeopathology*

Résumé: *Dans cet article, nous présentons un aperçu général sur les individus récemment exhumés des sambaquis fluviales (Vale do Ribeira, État de São Paulo, Brésil), en comparaison avec les rapports sur les collections de squelettes des constructeurs des sambaquis côtières hébergés dans différentes institutions. Nous commençons par une brève description des similitudes et des différences en ce qui concerne l'emplacement, la date, la dimension, de subsistance et de traits culturelles. Enfin, nous adressons caracteristiques palaeopatologiques. Nous concluons que la santé et le mode de vie de la population de le sambaqui fluvial Moraes se situe dans la variation observée dans les sambaquis côtières, sauf pour leur haute fréquence de la carie du cou. Ces résultats suggèrent principalement différences dans les stratégies de subsistance. Ces différences sont examinés dans Boyadjian* et al. *(cette volume) et d'apporter un éclairage sur le mode de vie des gens qui ont construit les sambaquis riveraines.*

Mots clés: *sambaquis fluvial, site d'implantation, affinités culturelles, subsistance, palaeopatologie*

INTRODUCTION

This paper aims at presenting a general palaeopathological overview on the recent findings of newly excavated riverine shellmounds located at the Vale do Ribeira (Southeastern part of the São Paulo State) in comparison with the more thoroughly studied coastal shellmounds scattered along most of the 8000 km of the Brazilian coast, based on already published data.

Coastal shellmounds or sambaquis are the best studied kind of archaeological site in Brazil. There are more than a thousand of these sites registered, dated from 8000 to about 1000 yBP (Gaspar, 1998). There are sambaquis along the rivers and along the sea coast. They constitute different prehistoric occupations sharing the shell platform among other characteristics.

Coastal sambaquis highly outnumber riverine sambaquis until now. There are more than 1000 coastal sites and only less than about 50 fluvial ones. Riverine sambaquis begun to be studied more systematically since the eighties, although their existence is known since the end of the XIX century (Barreto, 1988). They are known as "concheiros" or riverine sambaquis because these archaeological sites resemble coastal sambaquis in that there are also thick layers of mollusc shells, containing burials, habitation structures and remains giving clues about daily activities. They are circular or ovoid structures with about 500 to 2000m^2. The stratigraphic superposition of burial structures as well as the diversity of resources used suggest that the riverine sambaquis were used for at least some generations at a time.

One area that recently has received much attention for the study of riverine sambaquis is Vale do Ribeira, in São Paulo State. The Vale do Ribeira is a geographic and ecological transition between the highlands and the seashore and was inhabited since 9000 y BP. About 30 riverine sambaquis have been found in this region until now (Figuti *et al.*, 2004). Some of them were concomitantly active, indicating the existence of a net of joint structures. The proximity of most of the riverine sambaquis of the vale do Ribeira to the biggest river in the region suggests that the people who built these sites arrived at the valley coming from the coast and traveling up this river. This is supported by marine vestiges found in these riverine sites, such as shark teeth. (Barreto 1988).This type of archaeologial site, the so called riverine sambaquis or "concheiros", represents the oldest as well as the longest evidence of human occupation of this region (dating from 9000 to 1000 years BP), then there is the lithic tradition called Umbu (1200 yBP), and more recently (about 500 yBP) the ceramic tradition called Itararé. The Umbu lithic tradition is quite elaborate whereas the other stone artifacts, such as the projectile points are relatively fine and small. The exact meaning of the Umbu tradition in relation to the riverine sambaquis has yet to be determined. However, the Umbu tradition in the vale do Ribeira recalls the Mesolithic shellmounds of Muge, since there a similar lithic tradition has been found (Figuti *et al.*, 2004).

The oldest human sambaqui remains reaches back to about 9000 yBP and, more interesting, happened to be a riverine shellmoundbuilder from Vale do Ribeira, not a coastal dweller. Craniometric studies have shown great similarities of this individual (burial II from Capelinha I) with the Paleoindians, who inhabited the inland plateau earlier (13.000 to 8.000 yBP) than the coastal sambaqui groups appeared (Neves *et al.*, 2005). This suggests, a priori, a continental origin for the rivierine sambaqui groups with later cultural influence from coastal sambaquis. But, as discussed in the papers by Okumura *et al.* and Okumura & Neves in this issue, this hypothesis is not straightforward.

COMPARITIVE OVERVIEW BETWEEN RIVERINE AND COASTAL SAMBAQUIS

Coastal sites are usually located at estuarines. The mounds can be huge, especially in Southern regions, where they reach up to 30 m height. They are composed mainly of marine shells such as *Anomalocardia brasiliana* (Prous, 1992). Riverine sambaquis are found in different regions in Brazil, near rivers, are often located less than 100 km from the present sea shore and are much smaller in size and height than the coastal sambaquis. The shells used for construction are usually land snails, such as *Megalobulimus* (Barreto, 1988; Figuti *et al.*, 2004).

Subsistence at coastal sites is based on aquatic resources, is rich, and previsible (Figuti, 1993; Figuti & Kokler, 1996; DeMasi, 1999). Plants do poorly preserve and as such research on this part of the diet is still at its beginning (Schell-Ybert *et al.*, 2003; Boyadjian, 2007). At riverine sambaquis subsistence is mainly terrestrial, although some fish remains exist, especially in those sites located very near to rivers (Figuti *et al.*, 2004) The big size of the *Megalobulimus* mollusks returns up to 300gr flesh, so this species must have constituted a part of diet. The large quantities of young and small animals of very distinct species suggests opportunistic hunt in a part of the sites (Figuti *et al.*, 2004). Some of the riverine sambaquis are indeed very small, suggesting that people could have sometimes buried their deceased beloved ones while moving. No zooliths have been found at riverine sites (Barreto, 1988; Figuti *et al.*, 2004), but they are frequent in Southern coastal sites (Gaspar, 1998).

GEOGRAPHIC AND TEMPORAL SHIFTS OF RIVERINE SAMBAQUIS AT VALE DO RIBIERA

Dates and geographic localization of the riverine sambaquis located at Vale do Ribeira show that the oldest ones, dated to about 9.000 yBP, are restricted to a small region. After an interval of 1500 years, a great quantity of riverine sambaquis appear to be expanding geographic and demographically between 7000 and 3500 yBP. Finally, after a second interval of 1800 years the more recent riverine sambaquis, dated to 1700 to 1200 yBP, are again restricted to a relatively small area. Additionally, it appears that older sambaquis tend to have been used during longer times than more recent ones (Figuti *et al.*, 2004). This pattern raises the question if these expansion and retraction movements of riverine sambaquis, as well as the progressively shorter periods of occupation reflect ecological shifts.

Palynology and sea level studies indicate an approximate parallelism to the expansion and retraction movements of the riverine sambaquis at Vale do Ribeira (Bissa, 1998). Riverine sambaquis are clearly older than the oldest coastal sambaquis, whereas the most recent riverine shellmounds were contemporaneous with the Umbu tradition. This means 8000 years of riverine sambaqui tradition at this particular Valley.

Around 9000 years BP the neighboring inland plateau was dry and cold and on the other side the sea level was well below the level seen today, that is, about 5 meters, but already rising. The sea shore, thus, was wider, the riverine sambaquis located further away from the sea than today. Coastal sambaquis located at the beach at this time, were flooding and today must be under water. Was the Vale do Ribeira a refuge zone for groups under demographic and ecological pressure (Robrahn & Blasis, 1998)?

The riverine sambaquis dated to between 7000 to 3500 years BP, were built while the climate was hotter and drier in the plateau, at the same time the coast witnessed the transgressive period, which made the sea level rise 2 m above the level it shows today. So, considerable portions of the coastal plain were underwater. Maybe those environmental pressures in the neighborhood lead to the expansion of the riverine sambaquis to the north and west (Figuti *et al.*, 2004).

The last riverine sambaqui period (1700 a 1200 years BP) at the Vale do Ribeira occurred concomitantly with the expansion of the tropical forest at the plateau and the retreat of the sea level to the level seen today. The people of the riverine sambaquis seem restricted to the upper Ribeira in this period, which could have been triggered by the forthcoming archaic hunter-gatherers or the first agricultural cultures coming from the inland (Figuti *et al.*, 2004).

CULTURAL AFFINITIES BETWEEN RIVERINE AND COASTAL SAMBAQUIS

Although systematic studies are still lacking, it seems that the lithics, bone and shell industries are quite similar between riverine and coastal sambaquis. There are adornments made of animal canines, more funerary offerings accompaning child than adult burials, no clear burial pattern, burials with ochre, and fishing hooks at both types of sites suggesting cultural identity or proximity (Edwards *et al.*, 2001; Figuti *et al.*, 2004). This leads to the question about contact. In fact there are evidences of contact between coastal and fluvial sites at the southeastern region of Brazil: oysters, clams, ray spines, shark teeth and beeds made of marine shells found in riverine sambaquis (Figuti *et al.*, 2004). This suggests a coastal origin for the riverine sites. Maybe coastal sambaqui dwellers have used the Vale do Ribeira to reach the inland.

The nature of this contact is still in debate. Did inland populations migrate to the Vale do Ribeira adopting cultural habits from sambaqui populations? Did these contacts result in common offspring? This is what the morphological similarity between the first riverine sambaqui man (Capelinha II) and the Paleoindians suggest (Neves *et al.*, 2005). However, the cultural similarity between the riverine and coastal sambaquis is big (Figuti *et al.*, 2004), and there is no evidence of inland culture among the riverine sambaquis. Furhtermore, the oldest coastal sambaquis must be underwater.

Meanwhile there is no clear response to these questions, we report what is known until now about the way of life and health of riverine versus coastal sambaqui groups.

WAY OF LIFE AND HEALTH OF RIVERINE VERSUS COASTAL SAMBAQUI GROUPS

Osteologic and paleopathologic studies were carried out on eight coastal and one riverine collection of skeletonized human remains. Only one riverine site was included because the others yielded too small a sample until the present. The results presented below and in Figure 9.1 are taken from our previous studies where detailed descriptions about materials and methods can be found (Storto *et al.*, 1998; Okumura & Eggers, 2005; Scheel-Ybert *et al.*, 2003; Eggers *et al.*, 2003). The aim of this section of the paper is to contrast the health features of the people from the riverine sambaqui Moraes with those from coastal sites belonging to neighboring states.

Astonishingly, cribra orbitalia, considering it from the pinpoint perforation on, is quite high among most sambaquis, including the riverine Moraes, as is also unspecific periostitis (Figure 9.1). This suggests a surprisingly high pathogen load even for riverine populations, theoretically not so much exposed to unhealthy environments. An alternative explanation would be related to other stress factors such as inadequate diet and nutrition.

Although Moraes shows high frequencies of osteoarthritis, arms and legs seem almost equally affected, in the contrary to some of the coastal sites, such as Jabuticabeira II and Rio Comprido, with upper limbs more affected than lower limbs. This suggests that upper and lower limbs were equally engaged in physical activities among riverine dwellers, whereas coastal populations generally relied more heavily on the upper extremities for rowing, swimming and diving. In fact, the zooarchaeological studies in the riverine sites reveal that diet was based on a large range of terrestrial game, a fact that may explain a more generalized use of the body among the riverine shellmound builders.

Most sambaqui people, be they coastal or riverine, seem quite short, but they are in the range of extant and living Amerindians (Storto *et al.*, 1999). This is discussed as being the result of genetic characteristics and / or an adaptation to physiological stressors occurring especially in the weaning period (Dangour, 2001; Mendonça de Souza, 1995).

People from Moraes show much higher frequencies of caries in comparison to coastal sambaqui dwellers. Although the caries frequency for Moraes is well over the 10% "threshold" suggestive of agriculture (Turner, 1979), it does not mean that their subsistence necessarily was based on highly cariogenic and starch rich domesticated plants, since a part of them are root caries. Root caries originate more from mechanically impacted food residues, periodontitis (Reiker *et al.*, 1999), dental plaque (Fejerskov, 1997) and occasionally also from cariogenic diet and habits. More data on the diet of the people of Moraes is presented by Boyadjian *et al.* in this issue.

Contrary to the cold water hypothesis of auditory exostosis (Fabiani *et al.*, 1984) the people from Moraes, who were terrestrial hunters, show surprisingly high frequencies of auditory exostosis. As we have shown elsewhere, however, this might be due to effects of wind chill and cold atmospheric temperatures, even if contact of the ear canal with water was rather restricted (Okumura *et al.*, 2005/05, 2007a, 2007b).

Summing up, health and way of life of the people of the fluvial sambaqui Moraes lies in the range of distribution of paleopathologies observed among coastal sambaqui groups, except for their high neck caries frequency. These results suggest mainly differences in subsistence strategies. These differences are explored in Boyadjian *et al.* (this issue) and will shed new light on the way of life of the people who built the riverine sambaquis in the vale do Ribeira.

Acknowledgements

This work could not have been carried out without the help from the following institutions and people: Museu Arqueológico de Sambaqui MASJ, Museu Nacional do Rio de Janeiro, Museu de Arqueologia e Etnologia da USP, and their staff, Financial support: FAPESP (Grants Numbers: 04/11038-0, 1999/12684-2), Cecilia Petronilho, Célia Boyadjian, José Filippini, Ligia and Rafael Bartolomucci.

References

BARRETO, C.N.G.P. (1988) A ocupação pré-colonial do vale do Ribeira de Iguape, SP: sítios concheiros do médio curso. Dissertação de mestrado, FFLCH-USP.

BECK, A. (1970) Os sambaquis do Brasil meridional: litoral de Santa Catarina. Simpósio de Arqueologia Leste-Sul da América do Sul.

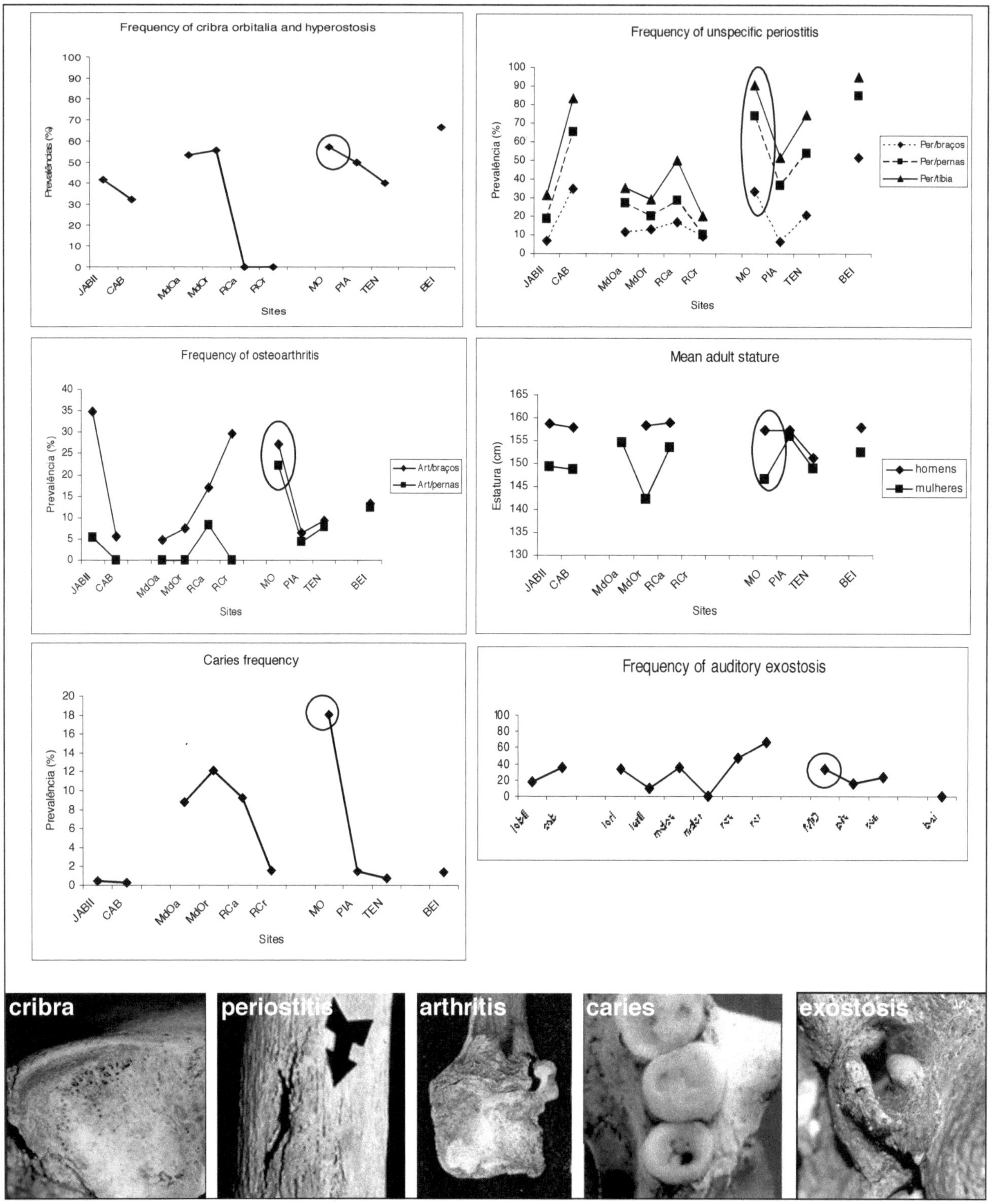

Fig. 9.1. Frequency of pathologies and mean stature in Brazilian sambaqui populations grouped according to decreasing latitude. The data from the riverine site Moraes (MO) is circled. JabII: Jabuticabeira II; CAB: Cabeçudas; MdO: Morro do Ouro; RC: Rio Comprido; MO: Moraes; PIA: Piaçaguera; TEN: Tenório; BEI: Beirada

BISSA, W. (1998) Reconstituição dos paleoambientes de uma planície holocênica no curso inferior dório Ribeira (município de Iguape, SP). Dissertação de mestrado, Depto Geografia da Univ São Paulo.

BOYADJIAN, C.H., EGGERS, S., REINHARD, K. (2007) Dental wash: a new method of estimating microfossil content in prehistoric teeth. Journal of Archaeological Sciences 34:1622-28.

DANGOUR, A.D. (2001) Growth of upper- and lower-body segments, in Patamona and Wapishana Amerindian children (cross-sectional data) ANNALS OF HUMAN BIOLOGY 28 (6): 649-663.

DeBLASIS, P.A.D., FISH, S.K., FISH, P.R., GASPAR, M.D. (1998) Some references for the discussion of complexity among the sambaqui moundbuilders from southern Brazil. Revista de Arqueologia Americana 9, México.

DeMASI, M.A. (1999) Pre-historic hunter gatherer mobility on the souythern Brazilian coast: Santa Catarina Island. PhD dissertation, Stanford University.

EDWARDS, H.G.M., FARWELL, D.W., de FARIA, D.L.A., MONTEIRO, A.M.F., AFONSO, M.C., De BLASIS, P., EGGERS, S. (2001) Raman spectroscopic study of 3000 year-old human skeletal remains from a sambaqui, Santa Catarina, Brazil. *J Raman Spectrosc*. 32, p. 17-22.

EGGERS, S., FILIPPINI, J., WESOLOWSKI, V., PETRONILHO, C.C., BOYADJIAN, C.H., BARTOLOMUCCI, R., GIARDINI, L.B. (2003). Bioarqueologia dos sambaquis fluviais do Vale do Ribeira. Anais do XII Congresso da Sociedade de Arqueologia Brasileira, [CD-rom].

FABIANI, M., BARBARA, M., FELIPO, R. (1984) External ear canal exostosis and aquatic sports. ORL Jounral of Otorhinolaryngology Relat Spec 46(3):159-164.

FEJERSKOV, O. (1997) Concepts of dental caries and their consequences for understanding the disease. Community Dentistry and Oral Epidemiology 25 (1): 1997.

FIGUTI, L. (1993) O homem pré-histórico, o molusco e o sambaqui: considerações sobre a subsistência dos povos sambaquieiros. Revista do Museu de Arqueologia e Etnologia USP 3:67-80.

FIGUTI, L. & KLOKLER, D. (1996) Resultados preliminares dos vestígios zooarqueológicos do sambaqui Espinheiros II (Joinville, SC). Revista do Museu de Arqueologia e Etnologia USP 6:169-188.

FIGUTI, L., MENDONÇA, C.A., PORSANI, J.L., ROCHA, E.B., DeBLASIS, P.A.D., EGGERS, S., BISSA, W.M. (2004) Investigações Arqueológicas e Geofísicas dos sambaquis fluviais do vale do Ribeira de Iguape, Estado de São Paulo. Relatório Final de Atividades de Projeto Temático (FAPESP no. 1999/12684-2).

FILIPPINI, J. (2004) *Biodistância entre sambaquieiros fluviais e costeiros: Uma abordagem não-métrica craniana entre três sítios fluviais do vale do Ribeira – SP (Moraes, Capelinha e Pavão XVI) e três costeiros do sul e sudeste do Brasil (Piaçaguera, Jabuticabeira II e Tenório*. MSc thesis, Universidade de São Paulo, Brazil.

GASPAR, M.D. (1998) Considerations of the Sambaquis of the Brazilian Coast. *Antiquity*. 72:277, p. 592-613.

GASPAR, M.D. (2000) Sambaqui: arqueologia do litoral brasileiro. Rio de Janeiro: Jorge Zahar.

MENDONÇA de SOUZA, S.M. (1995) Etresse, doença e adaptabilidade: estudo coparativo de dois grupos pré-históricos emperspectiva biocultural. Tese de doutorado da Escola nacional de Saúde Pública, Fundação Oswaldo Cruz.

NEVES, W.A., HUBBE, M., OKUMURA, M.M.M., GONZÁLEZ-JOSÉ, R., FIGUTI, L., EGGERS, S., De BLASIS, P.A. (2005) A new early holocene human skeleton from Brazil: implications for the settlement of the new world. *J Hum Evol*. 48(4):403-414.

OKUMURA, M.M.M., BOYADJIAN, C.H., EGGERS, S. (2005/2006) Análise da exostose do meato auditivo externo como um marcador de atividade aquática em restos esqueletais humanos da costa e do interior do Brasil. Revista do Museu de Arqueologia e Etnologia USP 15-16:181-197.

OKUMURA, M.M.M., BOYADJIAN, C.H., EGGERS, S. (2007a) Auditory exostosis in coastal prehistoric settlements in Brazil. Am J Physical Anthropology 132, (4): 558-567.

OKUMURA, M.M.M., BOYADJIAN, C.H., EGGERS, S. (2007b) Is cold water sufficient to trigger auditory exostosis? The Ear, Nose and Throat Journal 86(8):468-472.

OKUMURA, M.M.M., EGGERS, S. (2005) The People of Jaboticabeira II: reconstruction of the way of life in a Brazilian shellmound. *Homo*. 55:263-81.

PROUS, A. (1992) Arqueologia Brasileira. Brasília: Editora Universidade de Brasília.

REIKER, J., van der VELDEN, U., BARENDREGT, D.S., LOOS, B.G. (1999) A cross-sectional study into the prevalence of root caries in periodontal maintenance patients. Journal of Clinical Periodontology 26 (1): 26-32.

ROBRAHN-GONZALEZ, E.M. & DeBLASIS, P.A.D., (1998) Investigações arqueológicas no médio/baixo vale do Ribeira de Iguape, São Paulo. Ver Museu de Arqueologia e Etnologia (USP) 8: 57-69.

SCHEEL-YBERT, R., EGGERS, S. WESOLOWSKI, V., PETRONILHO, C.C., BOYADJIAN, C.H., DeBLASIS, P.A.D., BARBOSA-GUIMARÃES, M., GASPAR, M.D. (2003) Novas perspectivas na reconstituição do modo de vida dos sambaquieiros: uma abordagem multidisciplinar. Revista de Arqueologia 16:109-137.

STORTO, C., EGGERS, S., LAHR, M.M. (1999) "Estudo preliminar das paleopatologias da população do sambaqui de Jaboticabeira II". *Rev do Museu de Arqueologia e Etnologia USP*. 9:61-71.

TURNER, C.G. (1979) Dental anthropological indications for agriculture among the Jomon people of central Japan. American Journal of Physical Anthropology S1: 619-636.

Session C14

MODERN HUMANS ORIGINS IN EURASIA

TECHNICAL BEHAVIOR OF THE LEVANTINE AURIGNACIAN AT RAQEFET CAVE, MOUNT CARMEL, ISRAEL

György LENGYEL
Herman Ottó Museum, 28 Görgey, 3529 Miskolc, Hungary,
lengyelgu@freemail.hu

***Abstract**: The Aurignacian is characterized as bearer of modern human behavior. From lithic technological point of view this is expressed by a developed blade debitage and blade tool production. In the Levant, the Aurignacian lithic technology differs from the European view on Aurignacian. Levantine Aurignacian is characterized by a strong flake debitage and a weak blade debitage performed with hard hammer percussion technique. These features are more characteristic to the Middle Palaeolithic and the Middle to Upper Palaeolithic transitional industries than to any non-Aurignacian early Upper Palaeolithic culture in the Levant.*
***Keywords**: Levantine Aurignacian, lithic technology, technical behavior, Mount Carmel*

RAQEFET CAVE

Raqefet Cave is situated in the south-eastern side of Mount Carmel in Israel, on the left bank of wadi Raqefet, 230 m asl, approximately 50 m above the wadi bed (Olami 1984) (Figure 10.1). It is 50 m long with an area ca. 500 square meters. Eric Higgs of the Cambridge University and Tamar Noy of the Israel Museum conducted excavations at the site between 1970 and 1972, (Noy and Higgs 1971; Higgs *et al.* 1975). New excavations at the cave began in 2004 (Lengyel *et al.* 2005).

Studies on the lithic archaeological remains from the 1970-1972 stratigraphic units in squares B-G/18-23 assign Late Mousterian or Middle to Upper Palaeolithic transition (layers VIII-VI), indeterminate Early Upper Palaeolithic (layer IV), Levantine Aurignacian (layers IV, III and II), indeterminate Late Upper Palaeolithic (layer II in area), Late Kebaran (layer I), in squares J-M/24-28 Geometric Kebaran (layer VII) and Neolithic (I-IV) Late Natufian (layers IV-VI), and Bronze Age occupations all over the site (Higgs *et al.* 1975.; Lengyel 2003, 2005; Lengyel and Bocquentin 2005; Lengyel *et al.* 2005; Noy and Higgs 1971; Sarel 2004; Ziffer 1978a, 1978b).

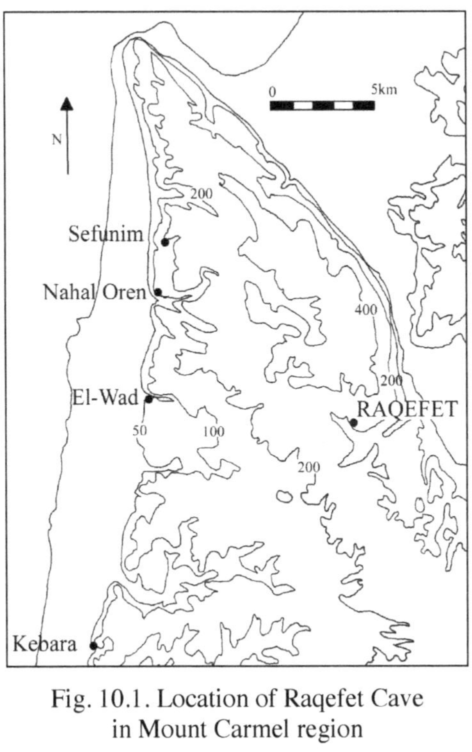

Fig. 10.1. Location of Raqefet Cave in Mount Carmel region

THE AURIGNACIAN ASSEMBLAGE OF RAQEFET

The majority of the Aurignacian assemblage derives from Layer III, and a few dozens of artifacts were found in the very base of Layer II, and the very top of Layer IV in single squares (Figure 10.2). Layer III is a 10-20 cm thick, silver/grey, slightly brecciated deposit. It is finely laminated by organic rich, ca 1-3 mm thick bands, horizontally running as irregular waves in section. The base of the layer in squares B-C/18-20 was reddish in color and varied in thickness from 1-5 cm. Layer II and IV are also laminated, although the black laminas are up to 1 cm in thickness. Charcoal samples of Layer III gave radiocarbon dates of 30540 ± 440 (RTT-4937) by AMS (Lengyel *et al.* 2006) and 33810 ± 1740 (I-6866) by decay counting method (Lengyel *et al.* 2005).

The lithic assemblage consists of mainly flakes and tools. As characteristic to Levantine Aurignacian assemblages, the number of blades and bladelets is a lot lower than that of the flakes (Table 10.1).

The high frequency of nosed, shouldered and carinated endscrapers, and typical Aurignacian retouch forms (Table 10.2) make Raqefet Aurignacian typologically similar to Ksar 'Aqil 8 and 7 in Lebanon (Dortch 1970, Bergman 1987), Hayonim D (Belfer-Cohen and Bar-Yosef 1981,), Kebara D (Garrod 1954, Ziffer 1978 a, b), and el-Wad D (Garrod and Bate 1937) in Israel.

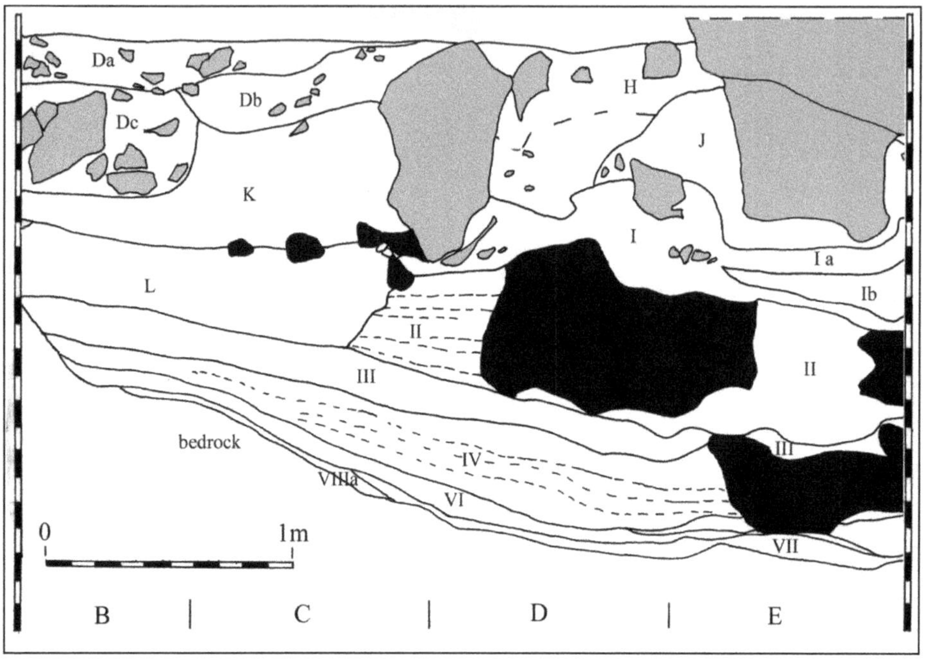

Fig. 10.2. Stratigraphic position of the Levantine Aurignacian layer in Raqefet

Tab. 10.1. Frequency of lithic products of the Aurignacian assemblage of Raqefet

flake	blade	bladelet	waste	CTE	tool	flake core	blade core	Bladelet core	pre-core	Total
1233	227	155	681	64	942	100	42	65	15	3524
35%	6,4%	4,4%	19,3%	1,8%	26,7%	2,8%	1,2%	1,8%	0,4%	100%

Tab. 10.2. Type list of the Aurignacian assemblage of Raqefet

Tool types		flake	blade	bladelet	waste	#	%
End-scrapers	simple	52	12		2	66	7
	on retouched blank	31	29			60	6,4
	ogival	7	6			13	1,4
	nosed	1	3			4	0,4
	thick nosed	41	8		4	53	5,6
	shouldered	3	1			4	0,4
	thick shouldered	26	3			29	3,1
	carinated	17	7		2	26	2,8
	double carinated	1				1	0,1
	circular	1				1	0,1
	thumbnail	1				1	0,1
	double	3	2			5	0,5
	Sub-total	***184***	***71***		***8***	***263***	***27,9***
Burins	carinated	8	1		1	10	1,1
	dihedral	9	6		1	16	1,7
	double dihedral	2				2	0,2
	on break	16	6	1	2	25	2,7
	on straight truncation	2	1		1	4	0,4
	on oblique truncation	3	4		1	8	0,8
	on concave truncation	3	2			5	0,5

Tool types		**flake**	**blade**	**bladelet**	**waste**	**#**	**%**
Burins	transversal	7	3		1	11	1,2
	flat	2	1			3	0,3
	multiple	3	3			6	0,6
	double angled	3	3			6	0,6
	Sub-total	***58***	***30***	***1***	***7***	***96***	***10,2***
Retouched items	continuous	18	15	5	1	39	4,2
	discontinuous	7	9	2		18	1,9
	partial	66	22	11	9	108	11,5
	inverse continuous			8		8	0,8
	continuously alternating	18	1	1		20	2,1
	discontinuously alternating	9		1		10	1
	partial alternating	11	2	1	2	16	1,7
	both edges continuous	3	3			4	0,6
	both edges partial	3	2			5	0,5
	strangulated blade		3			3	0,3
	alternate	2		1		3	0,3
	both edges mixed retouches	1	10	3		14	1,5
	Sub-total	***138***	***67***	***33***	***12***	***250***	***26,5***
Backed items	straight partially	6	2			8	0,8
	curved partially	1	1			2	0,2
	irregular		1			1	0,1
	Sub-total	***7***	***4***			***11***	***1,2***
Truncations	straight	17	5	2		24	2,5
	concave	4	1	2		7	0,7
	Sub-total	***21***	***6***	***4***		***31***	***3,3***
Points	el-Wad		3	1		4	0,4
Borers		29	3	1	3	36	3,8
Notches		108	19	5	17	146	15,8
Denticulates		34	10	3	6	53	5,6
Side-scrapers	simple	13	2			15	1,6
	double	2				2	0,2
	transversal	2				2	0,2
	concave	1				1	0,1
	convex	2	1			3	0,3
	dejete	3				3	0,3
	Sub-total	***23***	***3***			***26***	***2,8***
Divers	shouldered			3		3	0,3
	splintered tool	6			1	7	0,7
	raclette	4				4	0,4
	Sub-total	***10***		***3***	***1***	***14***	***1,4***
Composites	endscraper/burin	5	1			6	0,6
	endscraper/borer	1				1	0,1
	truncation/borer	1				1	0,1
	truncation/burin		1			1	0,1
	Sub-total	***7***	***2***			***9***	***0,9***
Grand Total		**619**	**218**	**51**	**54**	**942**	**100**

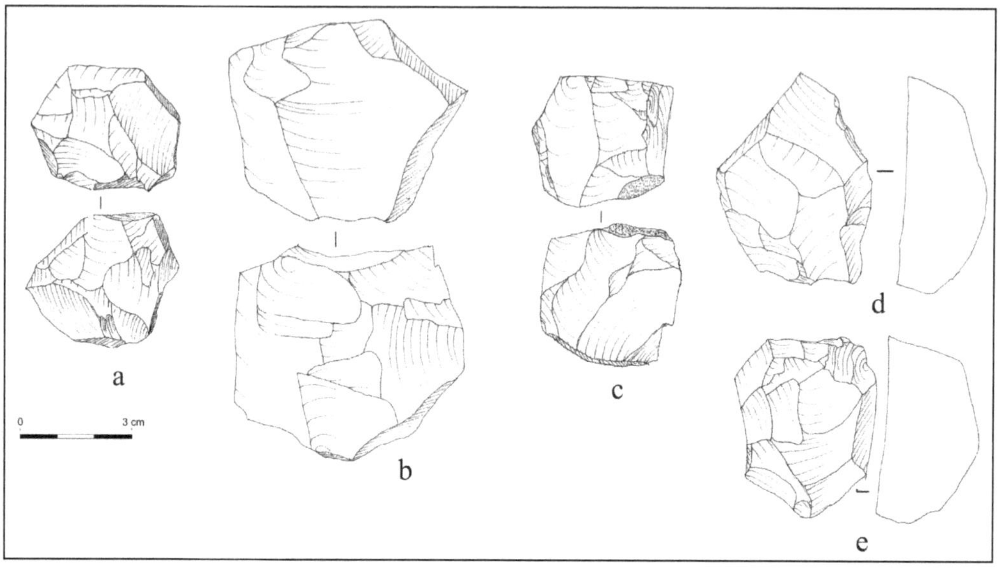

Fig. 10.3. Multiplatform flake cores (a-c) and thick flakes with multidirectional dorsal scars (d-e)

RAW MATERIALS

The raw materials of the Levantine Aurignacian of Raqefet mainly are of low and mediocre quality. These flints derive from south to the site in close distance, the area called Ramot Menashe in Lower-Middle Eocene geological formation consisting of chalk and chert. The small amount of good quality flints was collected mostly from north to the cave in Deir Hannah formation of Cenomanian age that consists of limestone, dolomite, marl, chalk and chert (Delage 2001).

FLAKE PRODUCTION

Most flake cores (n=76) have globular or cube shape with multiple striking platforms evidencing multiple orientation of the debitage (Figure 10.3, a-c). During multidirectional debitage the dimension of the core was intensively reduced from all surfaces. Flakes from these cores were systematically removed in order to keep the core in globular or cube shape. These flakes have steep distal termination (Figure 10.3, d-e). The non-unidirectional scar bearer flakes outnumber unidirectional scar bearer ones among thick items. Accordingly, this debitage floating and changing the debitage surface and the striking platform was designed for the production of thick flakes.

Flakes, usually thinner ones, were produced from single platform cores (n=18 cores). No remains of discoid debitage are found.

There are 42 flakes which were produced from flake's ventral face. This flaking modality is somewhat similar to the methods of Kombewa (Inizan *et al.* 1995) and Nahr Ibrahim (Solecki and Solecki 1970) which produce flakes with two ventral faces.

Basically, hard hammer technique was used to obtain flakes. The high frequency of thick butts (63% > 4 mm), presence of bulb (57%) and impact points (38%) on flakes illustrate that. Also, of 17 lateral fragments 9 specimens are typical Siret breaks, which are characteristic to the hard hammer percussion (Inizan *et al.* 1995).

Flake tools

A total of 31.4% of the flakes are tools and most of the tools were made of flakes, accounting for 65.7% of the tool kit. They are frequently cortical (54.6%). Included among tool blanks are 2 core tablets and 11 platform rejuvenating flakes. It is important to note that Janus flakes were not used as blank for tools.

Flakes are primarily blanks of endscrapers, retouched items, notches, and they dominate almost every tool type (Figure 10.4).

BLADE PRODUCTION

Blade core configuration

The blade production was based on mediocre quality flints. Of the 49 blade cores, 32 preserved a considerable part of the original shape of the initial raw material nodule. These illustrate that ca. 15-20 cm large nodular and tabular flints with longitudinal natural ridges were collected for blade production. This shape was ideal for beginning removing blades without front cresting (Figure 10.5, a). There is a lack of initial core shaping which is shown by the preponderance of cortical blades (53%) and cortical cores (78.6%).

The striking platforms of the cores were created by the removal of single flakes. Missing the appropriate raw

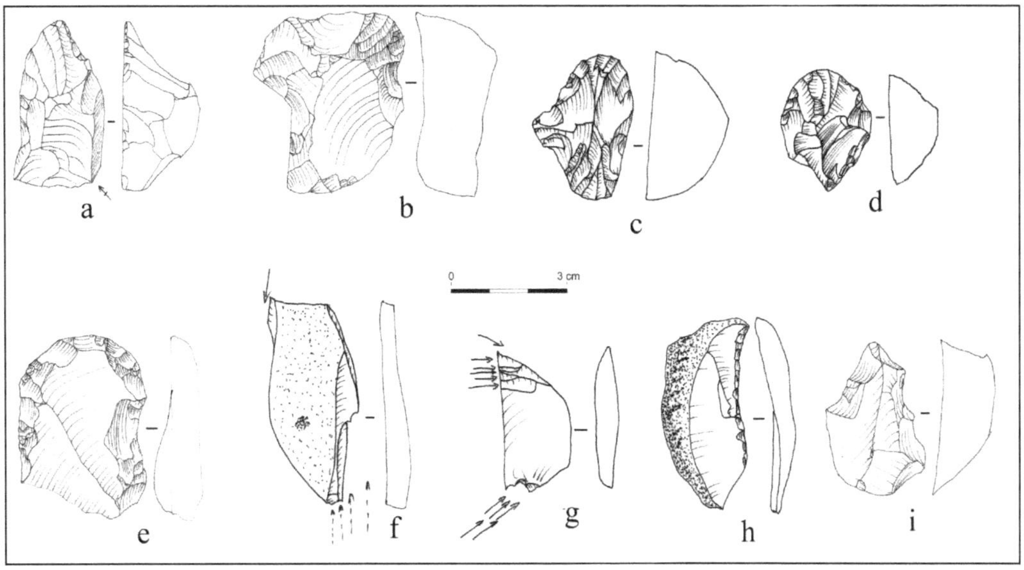

Fig. 10.4. Aurignacian type endscrapers (a-d), endscraper on retouched flake (e), burins (f-g), sidescraper (h), and notched flake (i)

material shape for blade production, which is rare, a new shape was given to the cores by flaking. One of the cases is, based on a partially refitted core, when after the striking platform creation a thick flake was detached from the right lateral side striking the striking platform, making the flint nodule considerably narrower.

Flake cores and large flakes also were used for making blade cores and subsequently blades. In the case of using large flakes, the striking platform was set up at the distal end, the front was crested (Figure 10.5, d), and the debitage took place on one of the edges.

Blade debitage

The striking platform of the cores remained plain during the debitage. Dihedral and faceted buts make up only 13%. The lack of overhang treatment (75.1%) and the strong presence of impact points (25.8%, compared to 38% frequency at flakes), illustrate that a large percent of the blades were removed by hard hammer technique.

Generally, the length and the thickness of the hard hammer detached blades are greater than those of the soft hammer removed ones. The largest blades (80-120 mm) were most often detached by hard hammer technique. Smaller blades (50-80 mm) were produced with both hard and soft hammer technique, while the smallest and finest blades (40-50 mm) were detached by soft hammer technique with high frequency. The largest blades were produced from narrow-fronted cores, and made often of a low quality flint which can be found in the largest size (ca. 30-50 cm) on the field. Contrary to this, the soft hammer technique detached finest blades often derive from wide debitage surfaced cores (only wide fronted cores have abraded overhangs) (Figure 10.5, b). Since a total of 73% of the blades and 80% of the blade cores have unidirectional scars the debitage was dominantly unidirectional.

Blade core maintenance

Common core maintenance was the striking platform rejuvenation by partial and total core tablet removals (n=35 and n=11, respectively). The debitage surface convexity was maintained by neo-cresting (18 neo-crest blades). The neo-cresting was habitually made at the distal part of the cores and only in rare cases performed along the entire length of the debitage surface. Hinge accidents on the blade cores' debitage surfaces were remedied by detaching the surface bearing hinge negative scars (n=17).

Blade core discard

The cores were discarded often after a few successful blade removals (n=13); after striking platform rejuvenation (n=1); and after hinge accidents (n=19). Also the useless blade cores, instead of discard, were used for producing flakes (n=3).

Blade tools

A total of 47.6% of the blades are tools, however blades make up only 23.1% of the tool kit. More than half of the blade tool blanks are cortical (54.7%). The tool blanks are characterized by straight profile. Included among tool blanks are 3 crest and 6 neo-crest blades.

Commonly, blades are blanks of all tool types and were used most frequently for end-scrapers. The finest blades are exclusive blanks of el-Wad points (Figure 10.5, f-g), while normal blades are those of endscrapers and retouched pieces (Figure 10.5, e, h), and the largest blades

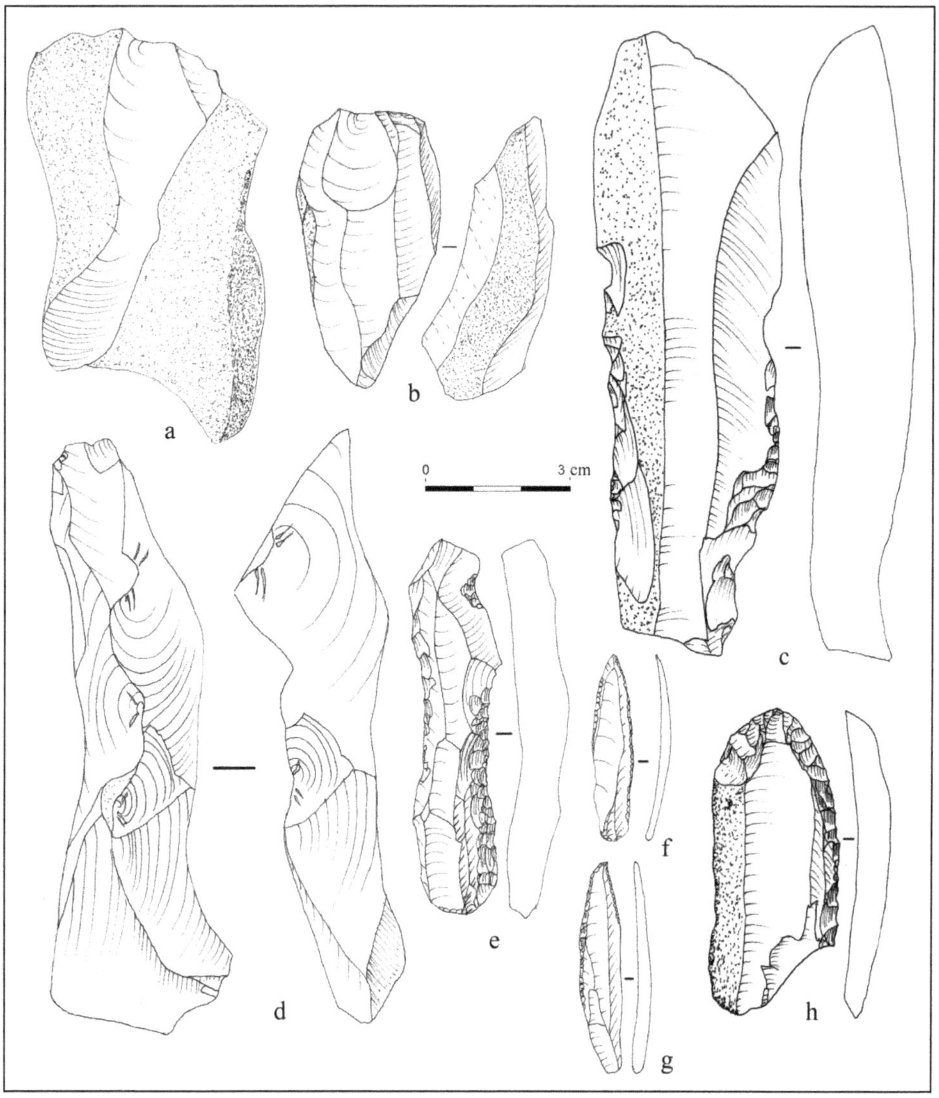

Fig. 10.5. Blade cores (a-b), denticulate blade (c), crest blade (d), strangulated blade (e), el-Wad points (f-g), endscraper on retouched blade (h)

are blanks for denticulates shaped by rough deep scaled retouch (Figure 10.5, c).

BLADELET PRODUCTION

Bladelet core configuration

The bladelet production is based on good quality raw material. Three types of blanks were used for bladelet cores. First are the small sized flint nodules (Figure 10.6, a). These, according to the preserved original surfaces on bladelet cores (n=8), were maximum 5 cm. The striking platforms of these cores were created by the removal of a single flake. After the striking platform, the lateral sides were flaked from the striking platform in order to create the debitage surface. The first bladelet removals were made along the ridges formed by flake removals. In rare cases, fully cortical bladelets were removed. The fronts of the cores made on small nodules are wide.

The second bladelet core blank type is the thick (14-37 mm) flakes (n=21) (Figure 10.6, b-c). These are frequently cortical (n=14). Their debitage surfaces were set up on the distal end or rarely on one of the edges of the flake. The bladelet debitage commenced without cresting on the unprepared flake edge. The core front is solely narrow.

The third bladelet core blank type is the exploited blade core. This is rare; only three cores have vestiges of prior blade production.

Bladelet debitage

The size of most bladelets (78.6%) ranges from 20 mm to 40 mm. Bladelets shorter than 20 mm are most likely the products of carinated and thick nosed and shouldered endscraper retouching (Figure 10.6, f-i, m).

The overhang abrasion is recorded in 41.8% of the cases and thin linear and punctiform butts (63%) are

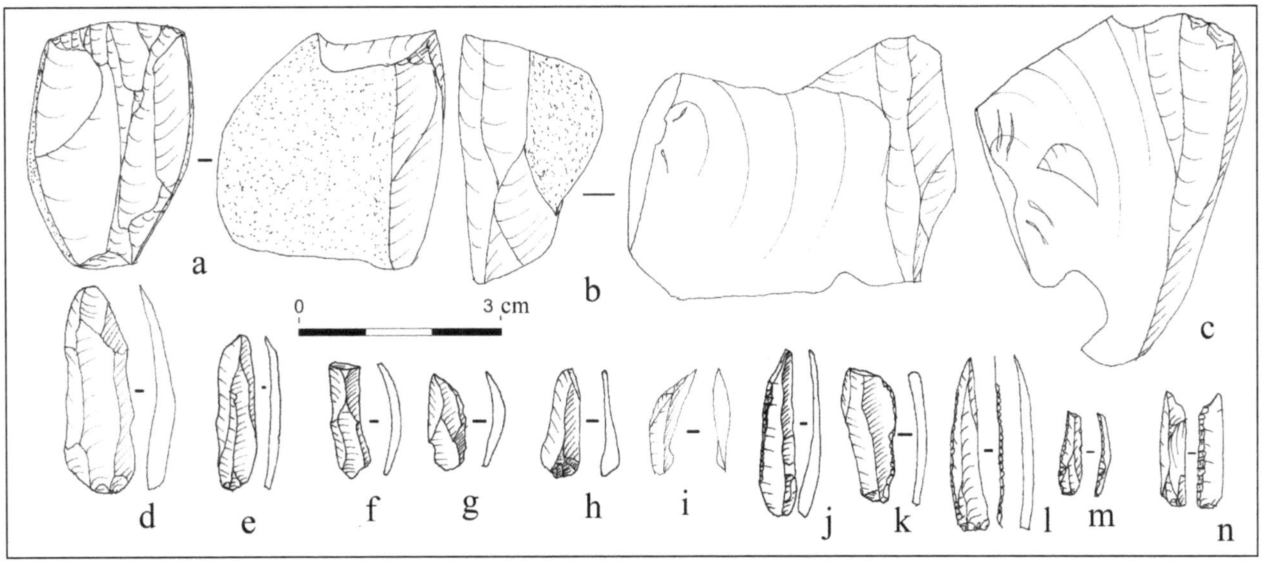

Fig. 10.6. Bladelet cores (a-c), bladelets (d-i), retouched bladelets (j-n)

characteristic to the bladelets. These data show that most of the bladelets were removed by soft hammer technique.

The overwhelming presence of unidirectional dorsal scar bearer bladelets (83%) over a few non-unidirectional bearer ones and the dominance of single striking platform cores mirror strong unidirectional bladelet debitage. Of the 65 bladelet cores, only 2 specimens have opposite striking platforms.

In profile, straight and curved shapes dominate. Twisted bladelets are relatively low in percent (21.3%) and are more abundant among 20-40 mm long products (27.4%).

Bladelet core maintenance

The high number of bladelet cores (n=65) compared to the low number of core tablets (n=4) and rejuvenating flakes (n=6) indicates uncommon use of striking platform rejuvenation. The debitage surface also was rarely corrected as indicated by the single neo-crest bladelet.

Bladelet core discard

The discard often occurred after a few ordinary removals. Hinge accidents led to core discard in 32.9% of the cases.

Bladelet tool blanks

A total of 24% of the bladelets are tools, which make up 5.4% of the tool kit. Only 13.7% of the bladelet tools are cortical. Bladelet blanks are frequently longer than 20 mm (n=47). There is a preference for using curved bladelets in tool kit. Most bladelet tools are simply retouched pieces (Figure 10.6, j-n).

AURIGNACIAN TECHNICAL BEHAVIOR AT RAQEFET CAVE

The Aurignacian technical behavior at Raqefet is based upon three main blank productions which are divided into, all together, seven sub-products.

Flake production

Thick flake debitage (>15 mm). This produces blanks with steep distal termination for thick endscrapers and notches-dencticulates. To produce thick flakes cube or globular-shaped cores with floating striking platform and debitage surfaces were used.

Thin flake debitage (<15 mm). This produces blanks for notches, burins and retouched tools. For producing thin flakes unidirectional debitage was used.

Blade production

Nodules for large blade and blade production are selected according to an appropriate shape for long and large blades which needed no or minimal investment in core preparation. The blade blank debitage is unidirectional, using consistently a single front and a single striking platform of the core. Rejected flake cores and thick flakes also are used for blade production.

Large blade debitage (80-120 mm). This produces blanks for denticulates from low quality raw materials. The cores are narrow fronted and the debitage is particularly done by using hard hammer technique.

Blade debitage (50-80 mm). This produces blanks for edge retouched tools and endscrapers. The cores, similarly to the former, are mostly narrow fronted. The debitage is done by using both hard and soft hammer techniques. There seems to be an advantage for the former technique.

Fine blade debitage (40-50 mm). This produces slightly curved blanks with converging lateral edges for el-Wad points. The cores are wide fronted and the detaching technique is soft hammer.

Bladelet production

The bladelet tool blank production is not the continuance of the blade debitage. The bladelet debitage does not plan to remove several series of bladelets from a single core. In the case of knapping accidents the cores can be easily replaced by new ones instead of rejuvenation.

Bladelet debitage (20-40 mm). This produces blanks for finely retouched tools. Small nodules were collected and thick flakes were produced for preparing bladelet cores. The bladelet core front habitually is narrow. The detaching technique is soft hammer.

Micro bladelet debitage (10-20 mm). This produces bladelets from thick Aurignacian type, carinated and nosed endscrapers with soft hammer technique. These blanks are rarely used in tool kits.

DISCUSSION

Levantine Aurignacian lithic technology besides Raqefet is known from Kebara (units I-II) and Hayonim (layer D) in Israel, and Ksar 'Aqil (layers 7-8) in Lebanon (Tostevin 2000; Williams 2003; Bergman 1987; Belfer-Cohen and Bar-Yosef 1981; Chazan 2001; Wiseman 1993), which can be characterized by the lack of using cresting, poor core shaping, unidirectional robust blade debitage with hard hammer technique, unidirectional fine blade debitage with soft hammer technique, intensive bladelet production, and flake production from multi platform cores. Another Aurignacian site in Israel, Sefunim (layer 8) is distinct with more intensive core preparation and use of bidirectional debitage (Williams 2003). Umm el Tlel in Syria also is introduced as Aurignacian site (Ploux and Soriano 2003). Since its assemblages published primarily contain remains of bladelet production, most tools a retouched bladelets, no Aurignacian type of endscrapers and blades, these cannot be classified Levantine Aurignacian. These are rather similar to the Late Ahmarian phenomena (Belfer-Cohen and Goring-Morris 2003).

In striking contrast to the Levantine Aurignacian, Early Upper Palaeolithic lithic industries lumped together under the term Early Ahmarian exhibit strong laminar production against flake production. In spite of their common cultural affinity to the Early Ahmarian, these sites demonstrate different approaches to flint knapping. Nahal Nizzana XIII in the Negev in Israel involves an intensive core decortication without front cresting and with unidirectional debitage (Davidson and Goring-Morris 2003). Boker A, also in the Negev, is characterized by the lack of core shaping and front cresting, and unidirectional debitage (Monigal 2003). Ücagizly (layers B-B4) in Turkey, Ksar 'Aqil (layers 17-16), Kebara (units III-IV), and Qafzeh (layer E) are characterized by front cresting, decortication and bi-directional opposite debitage (Kuhn *et al.* 2003; Ohnuma 1988; Tostevin 2000; Williams 2003; Bar-Yosef and Belfer-Cohen 2004). Ksar 'Aqil layers 13-11 which lacks typical Aurignacian features is characterized by core front cresting, decortication and unidirectional debitage. In Umm el Tlel, Ahmarian occupations beneath the so-called "Aurignacian" are also reported (Ploux and Soriano 2003). Bladelet production dominates these assemblages, which rather similar to the Late Ahmarian (Belfer-Cohen and Goring-Morris 2003). In Raqefet, beside the Aurignacian another Early Upper Palaeolithic assemblage of layer IV was recovered, which in the lack of diagnostic tool types in terms of archaeological culture called indeterminate (Lengyel 2005; Lengyel *et al.* 2005; Lengyel *et al.* 2006). The technology of this assemblage corresponds best to Boker A (Monigal 2003) and Ksar 'Aqil layers 13-11 (Bergman 1987). The main similarities between them are given by the predominant use of unidirectional laminar debitage and the lack of pronounced flake debitage. Dissimilarities between them are given by the frequent decortication and front cresting at Ksar 'Aqil 13-11.

CONCLUSION

In the Levant, Early Upper Palaeolithic non-Aurignacian industries represent a wide inter-site variety of knapping behavior, while the Aurignacian seems technologically uniform. This uniformity may signify a stern adherence to tradition in lithic production. Also characteristic to the Aurignacian is the use of several debitages while the Early Ahmarian is characterized by a single laminar debitage from blades to bladelets or the use of two debitages for blades and bladelets, respectively.

Characteristic elements of Aurignacian, such as the flake debitage and the hard hammer technique in blade production, are absent in Early Upper Palaeolithic non-Aurignacian industries in the Levant while common in the Middle Palaeolithic and in the Middle-Upper Palaeolithic transitional or Initial Upper Palaeolithic context (Kuhn *et al.* 2003, Meignen 2000, Meignen and Bar-Yosef 2002, Škrdla 2003, Tostevin 2000). Thus, in the Levant, the Aurignacian is distinct from other fully fledged Early Upper Palaeolithic culture. Although Levantine Aurignacian tool kits contain mostly typical Upper Palaeolithic types such as endscrapres and burins, a quite large percent is composed of Middle Palaeolithic types such as retouched flakes, sidescrapers, notches and denticulates. Strictly based upon the lithic products, it can be concluded that the Levantine Aurignacian technical behavior in lithic production is near to those in Middle

Palaeolithic and Transitional context and does not necessarily represent a clear Modern Human behavior.

References

BAR-YOSEF, O.; BELFER-COHEN, A. (2004) The Qafzeh Upper Paleolithic Assemblages: 70 Year Later. Eurasian Prehistory 2 (1). p. 145-180.

BELFER-COHEN, A.; BAR-YOSEF, O. (1981) The Aurignacian at Hayonim Cave. Paléorient 7/2. p.19-42.

BELFER-COHEN, A.; GORING-MORRIS, A.N. (2003) Current Issues in Levantine Upper Palaeolithic Research. In Goring-Morris, A.N.; Belfer-Cohen, A. eds. More Than Meets the Eye: Studies on Upper Palaeolithic Diversity in the Near East, Oxford, Oxbow: p.1-12.

BERGMAN, C.A. (1987) Ksar Akil, Lebanon. Technological and Typological Analysis of the Later Palaeolithic Levels of Ksar Akil. Vol II: Levels XIII-VI. BAR International Series 329.

CHAZAN, M. (2001) Bladelet production in the Aurignacian of Hayonim Cave, Israel. Palèorient 27/1. p. 81–88.

DAVIDZON, A.; GORING-MORRIS, A.N. (2003) Sealed in Stone: The Upper Palaeolithic Early Ahmarian Knapping Method in the Light of Refitting Studies at Nahal Nizzana XIII, Western Negev, Israel. Mitekufat Haeven. Journal of the Israel Prehistoric Society 33. p. 75-205.

DELAGE, C. (2001) Les ressources lithiques dans le nord d'Israël: La question des territoires d'approvisionnement natoufiens confrontée a l'hypothèse de leur sédentarité. Thèse de doctorat. Université de Paris I-Pantheon-Sorbonne. Vol 1-3.

DORTCH, C. (1970) The Late Aurignacian Industries of Levels 8–6 at Ksar 'Akil, Lebanon. Unpublished MA Thesis, UCL.

GARROD, D.A.E. (1954) Excavations at the Mugharet Kebara, Mount Carmel, 1931: the Aurignacian industries. Proceedings of the Prehistoric Society 20. p.155–192.

GARROD, D.A.E.; BATE, D.M.A. (1937) The Stone Age of Mount Carmel. Volume 1. Clarendon Press: Oxford.

HIGGS, E.S.; GARRARD, A.N.; NOY, T.; ZIFFER, D. (1975) Report on Excavation at Rakefet, Mount Carmel, Israel. Unpublished Raqefet Archives.

INIZAN, M.-L.; REDURON, M.; ROCHE, H.; TIXIER, J. (1995) Technologie de la Pierre taillé 4, Meudon: CREP, Nanterre.

KUHN, S.L.; STINER, M.C.; KERRY, K.W.; GÜLEC, E. (2003) The Early Upper Palaeolithic atÜçağizli Cave (Hatay, Turkey): Some Preliminary Results. In Goring-Morris, A.N.; Belfer-Cohen, A. eds. More Than Meets the Eye: Studies on Upper Palaeolithic Diversity in the Near East, Oxford, Oxbow: p. 106-117.

LENGYEL, G. (2003). Stone Artifacts as Imprints of Human Behavior: A Case Study of the Aurignacian of Raqefet Cave, Israel. In 4th International Conference of PhD Students, Humanities. University of Miskolc, Hungary. p.151-157.

LENGYEL, G. (2005). Lithic Technology of the Upper Palaeolithic and Epipalaeolithic of Raqefet Cave, Mount Carmel, Israel. Unpublished PhD thesis. University of Haifa.

LENGYEL, G.; BOCQUENTIN, F. (2005) Burials of Raqfet Cave in the context of the Late Natufian. Mitekufat Haeven. Journal of the Israel Prehistoric Society 35. p. 271-284.

LENGYEL G.; NADEL, D.; TSATSKIN, A.; BAR-OZ, G.; BAR-YOSEF-MAYER, D.; BEERI, R.; HERSHKOVITZ, I. (2005) Back to Raqefet Cave. Mitekufat Haeven. Journal of the Israel Prehistoric Society 35. p. 245-270.

LENGYEL, G.; BOARETTO, E.; FABRE, L.; RONEN, A. (2006) New AMS ^{14}C dates from the early Upper Paleolithic sequence of Raqefet Cave, Mount Carmel, Israel. Radiocarbon 48/2. p. 253-258.

MEIGNEN, L. (2000) Early Middle Palaeolithic Blade Technology in Southwestern Asia. Acta Anthropologica Sinica, supplement to Vol. 19. p. 170-180.

MEIGNEN, L., BAR-YOSEF, O. (2002) The lithic industries of the Middle and Upper Paleolithic of the Levant: Continuity or break? Archaeology, Ethnology and Anthropology of Eurasia 3 (11). p. 12-21.

MONIGAL, K. (2003) Technology, Economy, and Mobility at the Beginning of the Levantine Upper Palaeolithic. In Goring-Morris, A.N.; Belfer-Cohen, A. eds. More Than Meets the Eye: Studies on Upper Palaeolithic Diversity in the Near East, Oxford, Oxbow: p. 118-133.

NOY, T.; HIGGS, E.S. (1971) Raqefet Cave. Israel Exploration Journal 21. p. 225-226.

OHNUMA, K. (1988) Ksar 'Akil, Lebanon. A Technological Study of the Earlier Upper Palaeolithic Levels of Ksar 'Akil, vol. III. Levels XXV–XIV. BAR International Series 426, Oxford.

OLAMI, Y. (1984) Prehistoric Carmel. Haifa.

PLOUX, S.; SORIANO, S. (2003) Umm el Tlel, une sequence du Paléolithique supérieur en Syrie centrale. Industries lithiques et chronologie culturelle. Paléorient 29/2. p.5–34.

SAREL, J. (2004) The Middle-Upper Paleolithic Transition in Israel. Technological analysis. BAR. S1229.

SOLECKI R.L.; SOLECKI R.S. (1970) A new secondary flaking technique at the Nahr Ibrahim cave site. Bulletin du Musée de Beyrouth 23. p. 137-142.

ŠKRDLA, P. (2003) Comparison of Boker Tachtit and Stránská Skála MP/UP Transitional Industries. Mitekufat Haeven. Journal of The Israel Prehistoric Society 33. p. 37-73.

TOSTEVIN, G.B. (2000) Behavioral Change and Regional Variation across the Middle to Upper Paleolithic Transition. Unpublished PhD thesis. Harvard University, Cambridge.

WILLIAMS, J.K. (2003) Examining the Boundaries of the Levantine Aurignacian. Unpublished PhD thesis, Southern Methodist University.

WISEMAN, M.F. (1993) Lithic Blade Elements from the Southern Levant: A Diachronic View of Changing Technology and Design Process. Mitekufat Haeven. Journal of the Israel Prehistoric Society 25. p.13-102.

ZIFFER, D. (1978a) The use of technological and metric data in the study of four Levantine Aurignacian sites in the Mount Carmel region. A preliminary study. Paléorient 4/1. p. 71-94.

ZIFFER, D. (1978b) A Re-evaluation of the Upper Palaeolithic industries of the Kebara Cave and their place in the Aurignacian Culture of the Levant. Paléorient 4/2. p. 273-293.

Session C62

COASTAL GEOARCHAEOLOGY: THE RESEARCH OF SHELL MOUNDS

INTRODUCTION

organized by:
Marisa Coutinho AFONSO
(Museu de Arqueologia e Etnologia/Universidade de São Paulo, Brasil)

Geoff BAILEY
(Dept. of Archaeology, University of York, UK)

Our aim in this session was to present new research on the shell mounds from Brazil and to place that research into a wider comparative framework, including new geoarchaeological and bioarchaeological work on the shell mounds of Northwest Europe, including the classic sites of the Tagus Valley (Portugal) and the Danish Ertebølle, Central and South Americas and Australia. In particular we sought to bring together a range of contributions that assessed the variety of functional, environmental, social and symbolic factors that contribute to the formation of shell mounds, presented varied methodologies for their investigation, and discussed the challenges and possibilities of investigating use of coastal resources in earlier periods of lowered sea level.

Nine papers were presented at session C62 by: **José Rolão** and Tim Van-der-Schriek (The Muge Shell Mounds of Portugal: Cultural and Environmental Factors in Mound Formation); **Nina Nielsen** (Danish Stone Age Shell Middens and applied Geoarchaeological and Bioarchaeological Methods); **Geoff Bailey** (When is a mound not a mound? Shell deposits, palimpsests and the dynamics of archaeological site visibility); **Peter Hiscock** and Patrick Faulkner (Australian shell mounds, coastal economies, and Aboriginal social constructions); **Ignacio Clemente Conte**, Ermengol Gassiol Ballbè and Leonardo Lechado Ríos (Shell mounds of the Atlantic Coast of Nicaragua: something more than middens); **Manoel Gonzalez** and Sandra Nami Amenomori (The occupation of prehistoric fisher-gatherer groups in an insular landscape at the coast of São Paulo State, Brazil); **Flávio Rizzi Calippo** (The underwater shell mounds of Cananéia, Southern coast of São Paulo, Brazil); **Marisa Coutinho Afonso** and Laércio Loiola Brochier (Geo-archaeological Investigations of Shell Mounds, Southern Brazil); **Jordi Estévez Escalera** from the Devara Research Project Team and Ernesto Piana (Microstratigraphy of shell middens of Tierra del Fuego).

Rolão and **Van-Der-Schriek** discussed the cultural and environmental factors in mound formation of the Muge Shell Mounds of Portugal. The shell mounds of the Lower Tagus valley are one of the largest and most distinctive groups of Mesolithic shell mounds in Europe. They clearly represent middens in the conventional sense with plentiful remains of artefacts, animal bones and domestic features, but some were also used as burial mounds, the most notable example being the mound of Armoreira, with a reported 300 burials. They discussed the function and significance of the mounds in the light of the most recent archaeological investigations and examined the impact of changes in the availability and distribution of molluscan habitats on the growth of the mounds, using new geomorphological and palaeoenvironmental investigations of valley floor evolution in the Lower Tagus catchment.

Nielsen discussed the Danish Stone Age shell middens, which appear as oblong shell heaps along the prehistoric coast line. Food remains, artefacts and features suggest that daily activities took place at most Mesolithic middens, while the Neolithic middens may have had a more specialised function. Selected examples of how geoarchaeological and bioarchaeological analyses can contribute to the understanding of shell middens and the surrounding environment were presented.

Bailey explored a range of field data and models of site formation, taking account of geological, environmental and cultural factors that may favour mounding in some cases but not in others. Larger mounds must have become prominent and recognised features of the landscapes, which acquired significance for their makers that may have been quite different at different stages in the history of mound formation. These characteristics pose a number of challenges to interpretation, including the demographic and social interpretation of mounding, the reduced resolution that usually follows from successive and repeated deposition and use on the same surface, the identification of the symbolic significance that the mounds may have had for the people who lived on and around them. The presentation included the results of a new project in the southern Red Sea (Farasan Islands), which is studying a rich record of mid-Holocene shell mounds and exploring underwater for traces of shorelines and associated archaeology formed at earlier periods of sea level.

Hiscock and **Faulkner** reviewed archaeological evidence for the formation of the shell mounds found across an area of tropical northern Australia and examined models of the economies and social systems they might reveal. Similarities in the spatial and temporal patterns of these structures exist between disparate regions, revealing distinctive economic and political forager systems operating along the northern coast from more than 2.000 years ago until about 800-600 years ago, as Aboriginal people intensively exploited large beds of the *Anadara* mollusc. Formation of anthropogenic shell mounds in northern coastal Australia reflects an intricate relationship between Aboriginal foraging economy and the environment; neither the prehistoric economy nor its environmental setting survived until the historic period. This raises interesting challenges about the use of ethnographic information for reconstruction of the economy and social system of prehistoric shell mound builders, since it can be shown that the mounds have a quite specific social and symbolic role in the way of life of the present-day people that may be quite different from the prehistoric situation.

Clemente Conte, **Gassiot Ballbè** and **Lechado Ríos** presented the shell mounds of the Atlantic Coast of Nicaragua and the results of the extensive excavation of a shell midden in the Karoline site, in the central Caribbean coast of Nicaragua. The results of the fieldwork and the later analyses of the archaeological remains allow them to reconstruct aspects of social life in this village (occupied between 450 cal BC until 350cal AD), the use of domestic space inside and surrounding a hut, and the process of formation of the shell midden close to this habitation area. More than a simple midden area, the shell mound is a place where day-to-day activities were carried out, and the internal stratification shows a substantial accumulation of shell midden deposit, with a constant remodelling of the internal work spaces.

Gonzalez and **Amenomori** discussed the occupation of prehistoric fisher-gatherer groups in Southeast Brazil, and the analysis of eight archaeological sites on the coast of São Paulo State located on islands and near the shore. The groups who lived in that region shared the same landscape perception: a place that could offer shelter and protection. This symbolic meaning was important in the choice of location and even when the new area was not surrounded by water, the apparent aspect of an insular landscape given by the shape of a mound created a sense of safety for their occupants.

Calippo discussed the underwater shell mounds on the southern coast of Brazil. Among the Cananéia mangroves, on the coast of São Paulo State, eight shell mounds were located and showed a strong correlation with the sea-level related changes that occurred after the last regressive maximum. This evidence indicates that completely submerged shell mounds may be preserved and that sites related to earlier sea level are not always destroyed or submerged.

Afonso and **Brochier** discussed geoarchaeological investigations at shell mounds in Southern Brazil. New research on shell mounds, involving zooarchaeological, geoarchaeological, bioanthropological and chronological studies, carried out in the north and south parts of Santa Catarina State coast indicated long-duration occupations by people specialized in exploitation of sea resources (fish and shellfish), using shells to build the sites. Detailed researches were carried out at three different types of locations: large shell mounds visible as landmarks, small ones, and other types of coastal sites, displaying a very complex occupation pattern. This paper presented some sites in different areas as examples of the field methodology applied to shell mounds and data interpretation.

Estevez presented the results that the Devara Research Project Team and **Piana** have obtained from research on the shell middens of Tierra del Fuego (Argentina). This Argentinean-Spanish research project was the outcome of a systematic strategy for recovering the archaeological record of fisher-hunter-gatherer shell middens of the Beagle Channel. They presented the approach that has allowed for the identification of sequential episodes of occupation and for the reconstruction of a fine-grained sequence of site formation. Two sites (Tunel VII and Lanashuaia) were discussed to illustrate the advantages of this methodological approach for understanding shell midden formation processes. The approach relies on a well-designed recording system that can incorporate the spatial information combined with the results of micro-morphological analysis, bone and stone material refitting and biological remains.

SHELL MOUNDS, PALIMPSESTS AND THE DYNAMICS OF ARCHAEOLOGICAL SITE VISIBILITY

Geoff BAILEY
Department of Archaeology, University of York, York, Y01 7EP, UK

Abstract: *Shell mounds show a continuum of shapes and sizes, from the conical or dome-shaped mounds of tropical regions such as Brazil and northern Australia to small shell scatters and coastal deposits in which shells are entirely absent. Shell middens of any description are rare before the establishment of modern sea level, suggesting recent intensification of marine economies, but many now suspect that this may be a bias resulting from loss or submergence of earlier material formed at lowered sea level. This presentation examines models of site formation in relation to geological, environmental and cultural factors that affect the preservation of shell mounds in the archaeological record and their visibility as recognisable features in the landscape.*
Keywords: *Marine molluscs, Shell mounds, Shorelines, Submerged landscapes*

THE GENERAL PROBLEM

This paper examines some of the geological and environmental factors that affect the archaeological visibility of shell mounds and how these in their turn influence the social and economic interpretation of coastal economies. The underlying question is the origins and development of marine-oriented economies, generally considered to be a relatively late phenomenon in the wider sweep of human evolution. Shell mounds are a major archaeological indicator of such economies and the conventional assumption is that the size and distribution of these deposits correlates with the intensity of interest in marine resources. Here I consider two sorts of geoarchaeological factors that complicate this assessment. The first is the micro scale problem of what determines mounding of shells to form visible deposits rather than the scattering of shells in less visible form. As I shall demonstrate, there are significant geological and ecological variables that contribute to mound formation, although cultural factors are also involved. The second is the macro scale problem of sea-level change. Shell middens of any sort are relatively rare before modern sea level was established about 7000 to 6000 years ago and larger shell mounds are unknown before that period. That fact has generally been seen as evidence for an intensification of marine economies in the early to mid-Holocene. However, there is a growing belief that this process of intensification may be more apparent than real, due to the loss of earlier coastal evidence by inundation during sea level rise (Bailey 2004, Bailey & Milner 2002, Bailey & Craighead 2003, Erlandson 2001). That suspicion has increasingly focused attention on the need to explore the submerged landscape associated with lower sea levels. Sea levels have been at least 40 m lower than the present throughout most of human existence, and for shorter periods much lower (Figure 11.1). These sea-level changes have had obvious effects on the survival or visibility of coastal archaeological material formed at lower sea level. They have also had more subtle effects that could have exerted an equally powerful influence on human palaeogeography and the visibility of marine economies, including lateral shifts of shoreline position during periods of rapid sea-level change, and changes in shoreline ecology within the postglacial and at other periods. It follows that a substantial part of the Pleistocene record is missing, submerged or otherwise distorted by sea-level change, and the search is now on for traces of that submerged landscape and its associated archaeology.

MODELS OF MOUND FORMATION

Do shell mounds necessarily form wherever people have engaged in gathering of large quantities of molluscs? Here, I shall examine this question on the working assumption that shell middens are incremental structures, the cumulative product over time (years to centuries) of multiple episodes of deposition. It follows that the visibility of shell midden deposits is the result of multiple episodes of shell deposition superimposed one on the other in the same location. But this necessarily comes at a cost, which is that the resulting deposit is likely to be a palimpsest — a mixing together and partial removal of successive activities that are difficult to resolve into their individual constituent episodes (Bailey 2007) — and quite possibly a site that changed its function and its significance for the people who used it at different periods in the history of the accumulation. Conversely, a single episode of shell deposition, for example a single family meal or a single season's use of a particular campsite, may produce a layer of shells that is archaeologically invisible or nearly so, and the same is likely to be true of multiple episodes of deposition, each of which is located at a different spot in the wider landscape. From this flows a paradox. The paradox is that archaeologically visible deposits, especially large ones, are likely to be complex structures with a variable history of use and significance that may be difficult if not impossible to untangle. Conversely, sites that can be associated with specific and identifiable functions, purposes and activities are likely to be archaeologically invisible, rare, or confined to recent periods where processes of deformation have had little time as yet to erase or obscure the evidence.

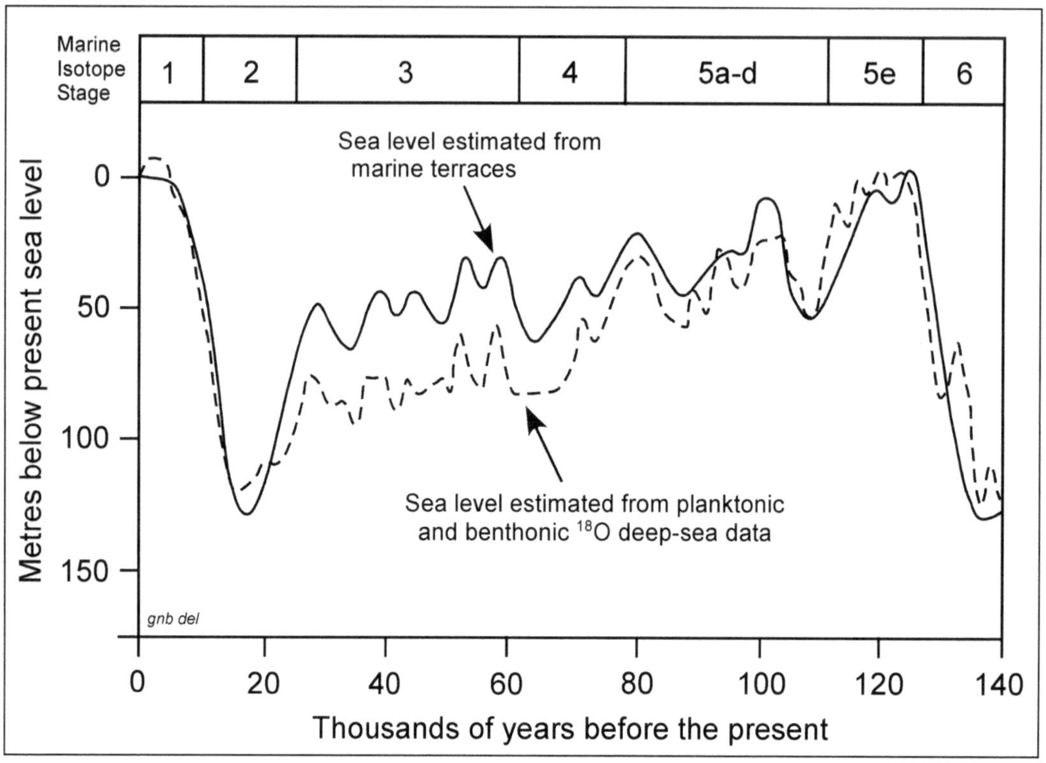

Fig. 11.1. Sea level change during the Last Glacial–Interglacial cycle. Data from Chappell & Shackleton 1986, Lambeck & Chappell 2001, Shackleton 1987, Siddall *et al.* 2003, Van Andel 1989). A similar cycle of sea-level change can be extended back to about 1 million years ago. Before that, the amplitude of variation was lower but the resulting pattern remains one in which sea levels lower than the present were probably the norm

This then leads on to the question of what accounts for repeated deposition of shells or other material evidence in the same location. One obvious precondition for the visibility of shell middens is the local availability and abundance of marine molluscs, and the persistence of the mollusc supply in the face of more or less subtle changes in sea-level, sediment supply and shoreline ecology, all of which can lead both to changes in species composition or abundance, and also to changes in the location of shell beds and the optimum locations for the collection and disposal of the shells. Generally speaking, river estuaries with shallow intertidal conditions are the most productive in ecological terms for marine molluscs but they are also the most sensitive to changes in ecology and spatial configuration. Conversely, steep rocky shorelines are much less sensitive to change, but equally they are less productive of molluscs, while shallow intertidal rock platforms may produce intermediate conditions.

A second set of factors has to do with the behaviour of shellgatherers and the constraints on shell disposal. One set of constraints results from automatic or unconscious factors: proximity to shell beds, or locational constraints tied to specific micro-environmental factors such as shelter, a shady tree or a water supply, where the repeated deposition of shells in the same location is the by-product of some other factor. Another set of constraints results from conscious decisions about shell disposal, and these may be motivated by practical reasons of rubbish disposal, or by ritual factors such as human burial, or 'increase' ceremonies, where shells are repeatedly piled in the same place in order to symbolise the significance of the rites associated with interment or other beliefs.

WEIPA: NORTHERN AUSTRALIA

Shell mounds, often of impressive dimensions, are widely distributed in Australia, especially in the tropical north. One of the largest, longest known and best studied group of sites is at Weipa in northern Queensland (Figure 11.2), where over 500 shell mounds of varying size have been recorded along the banks of the rivers that flow into Albatross Bay, the largest inlet on the west coast of the Cape York Peninsula (Bailey 1977, Bailey *et al.* 1994). The sites as a whole have a total time depth of about 2700 radiocarbon years, extending up to about the time of European contact. Shells form the main visible constituent of the deposits, and the dominant molluscan species is the bivalve *Anadara granosa*, which is typically found in large quantities on intertidal mudflats adjacent to mangrove lined shorelines and river banks. Only limited excavations have been undertaken at Weipa, but these demonstrate the presence of charcoal lenses, stone and bone artefacts and the bone remains of fish and marsupials. Analysis of shell quantities and rates of accumulation suggests that the amount of shell recorded in the deposits is consistent with a subsistence pattern in

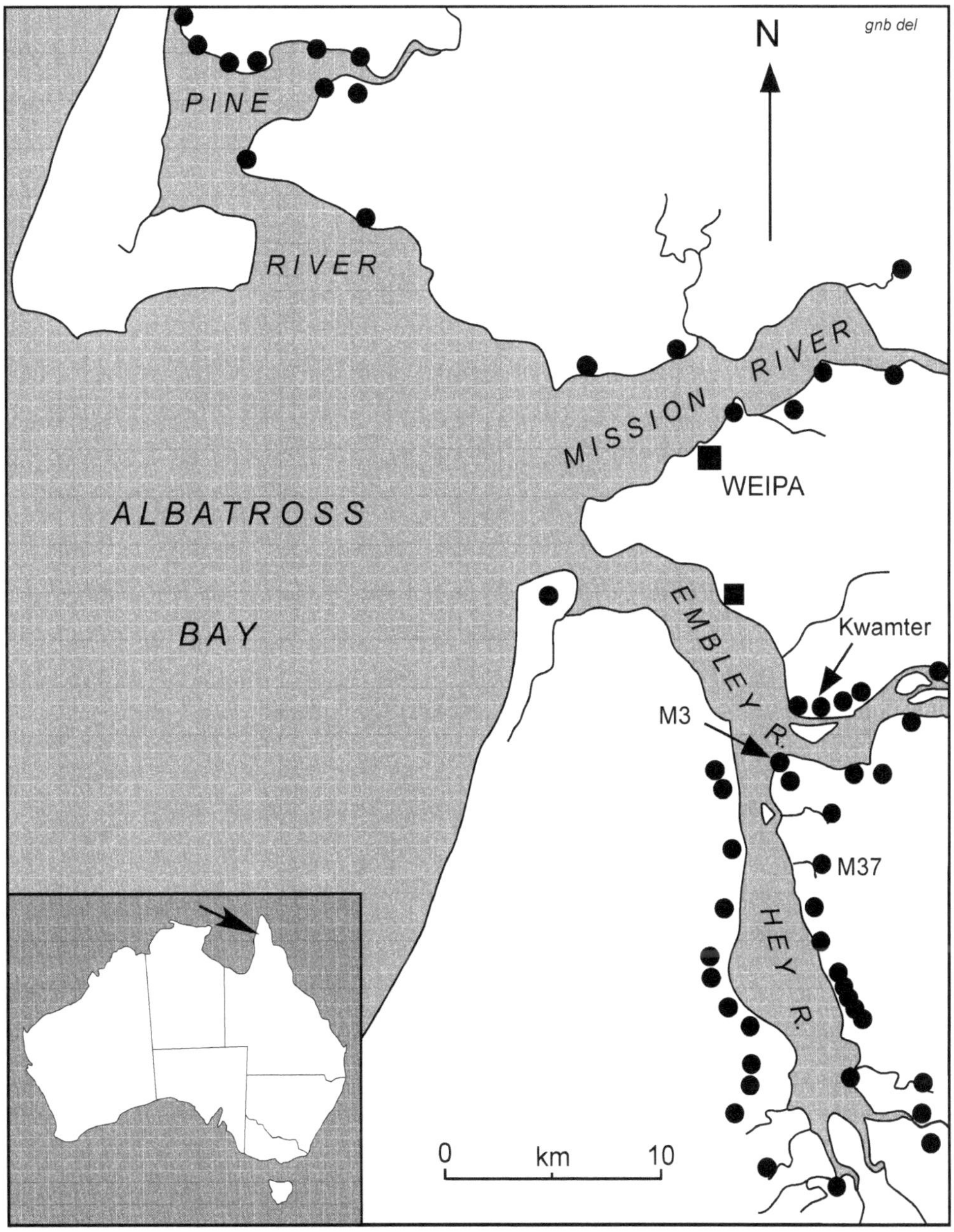

Fig. 11.2. Location map of Weipa shell mounds. Each circle on the map represents a cluster of shell mounds

which shells provided a relatively small contribution to the total food supply (cf. Meehan 1982). There have been repeated suggestions that the deposits cannot be the result of human activity and must have resulted from a variety of natural processes including storms and the nest-building activities of birds, but detailed analysis of site composition and location has eliminated these hypotheses and confirmed the essentially cultural origin of the shell mounds (Bailey 1993, 1994, Bailey *et al*. 1994).

The region is one of marked seasonal change of climate, with a dry season when the low-lying open ground of salt flats and sedge plains immediately behind the mangrove fringe is dry, and a wet season when these low-lying areas are flooded by rain and high tides. The mounds are variable in size and in micro-location. The largest mounds are over 10 m high with a maximum volume of 10.000 m^3, but the majority are much smaller deposits, and some are scarcely more than surface scatters of shell. Many sites are located on seasonally waterlogged ground in the open, and others are on dry sand ridges or in woodland on dry ground on the edge of the sandstone plateau. The shell mounds tend to form clusters with middens of different sizes in quite close proximity but in different microenvironments, some on dry ground, others on seasonally waterlogged ground, some in the open and others within woodland. The biggest mounds are usually quite irregular in shape and typically have a gentle ascent on at least one side with a small living area on the summit, clearly indicating that they are composite structures resulting from the long-term coalescence of smaller deposits.

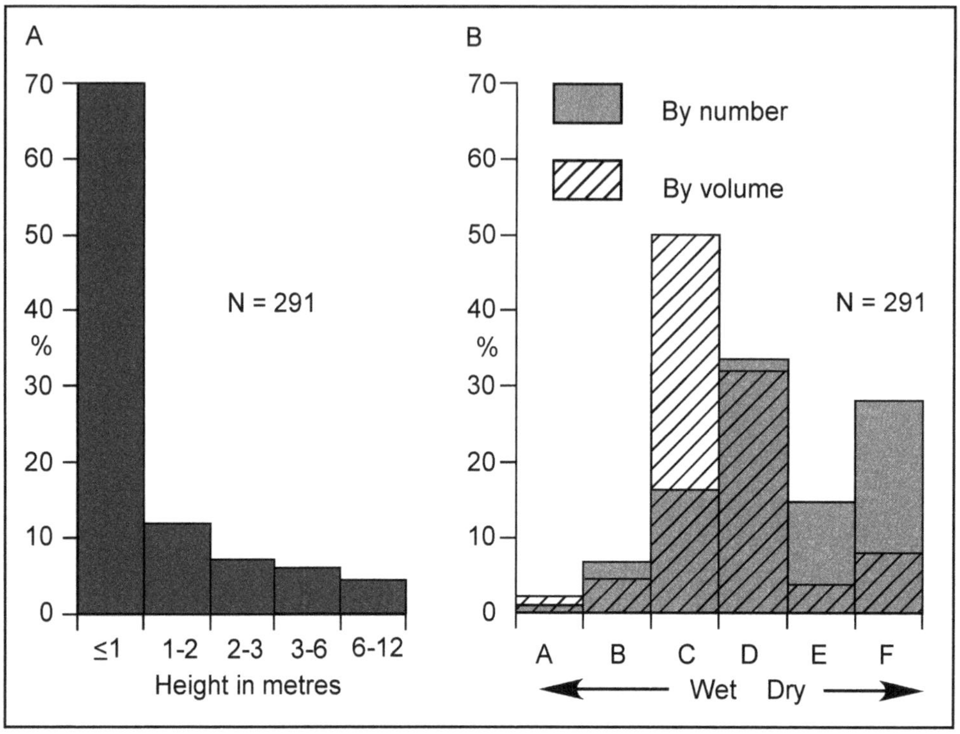

Fig. 11.3. Frequency distribution of Weipa shell mounds by mound height (left) and (right) by number and shell volume in relation to different types of substrate and microenvironment. A: Mangrove; B: Salt flat; C: Open Sedge Plains; D: Sand ridge (chenier) on salt flat or sedge plain; E: Edge of sandstone plateau; F: Sandstone plateau within woodland

This pattern can be explained in terms of four different hypotheses:

1. Self-selection for dry surfaces in a seasonally flooded landscape. At the end of the wet season and the early part of the dry season while low-lying ground near the river channel is still very wet, there would be a high premium on occupation close to the river on low-lying open ground both to get as close as possible to the food supplies in the river and to move out of tree cover onto open ground so as to avoid insect pests. This in its turn would lead to a strong preference for use of dry ground and hence for re-use of pre-existing shell surfaces raised above the wet ground, resulting in superimposition and upward growth of shell mounds. Conversely during the dry season, when the seasonally flooded low-lying ground is dried out, there would be less incentive or need to use mounded shell surfaces, and greater dispersal of the population across the landscape and in the hinterland. During the stormiest part of the wet season, preferences for camp sites would be on the higher drier ground within woodland cover to provide protection from rains and storms, and again there would be weaker tendencies to re-use previously accumulated shell surfaces.

2. Self-selection in relation to food-bearing trees and shrubs. The argument here is that these plants tend to find especially attractive conditions for growth in the shell matrix of midden deposits, especially if their seeds have been dropped there in previous episodes of food consumption, and that the presence of such plant foods reinforces the attractions of that particular spot for subsequent human occupation (Cribb 1986).

3. Intensification over time, with the small sites occupied early on in the sequence and the larger ones occupied later, presumably in relation to growth in the human population, expansion in the mollusc habitat, or both factors working together.

4. Social aggregation. On this hypothesis the mounds of different sizes are broadly contemporaneous as in hypotheses 1 and 2, but the combination of sites represents occupation by large temporary gatherings of family groups for purposes of ritual ceremonial and exchange, perhaps focused on shellfood gathering, consumption and shell disposal (Morrison 2003).

These hypotheses are not mutually exclusive, of course, and it may be that more than one factor is at work in a mutually reinforcing process of mound accumulation. As a partial test of these hypotheses we can look at the relationship between mound size and micro-location. The results shows that the great majority of the mounds are less than 1 m thick and show a bias to dry woodland locations, and that in terms of shell volume, the great bulk of shell is associated with seasonally waterlogged ground (Figure 11.3). This effect is even more marked if we look only at mounds taller than 3 m, with the tallest mounds

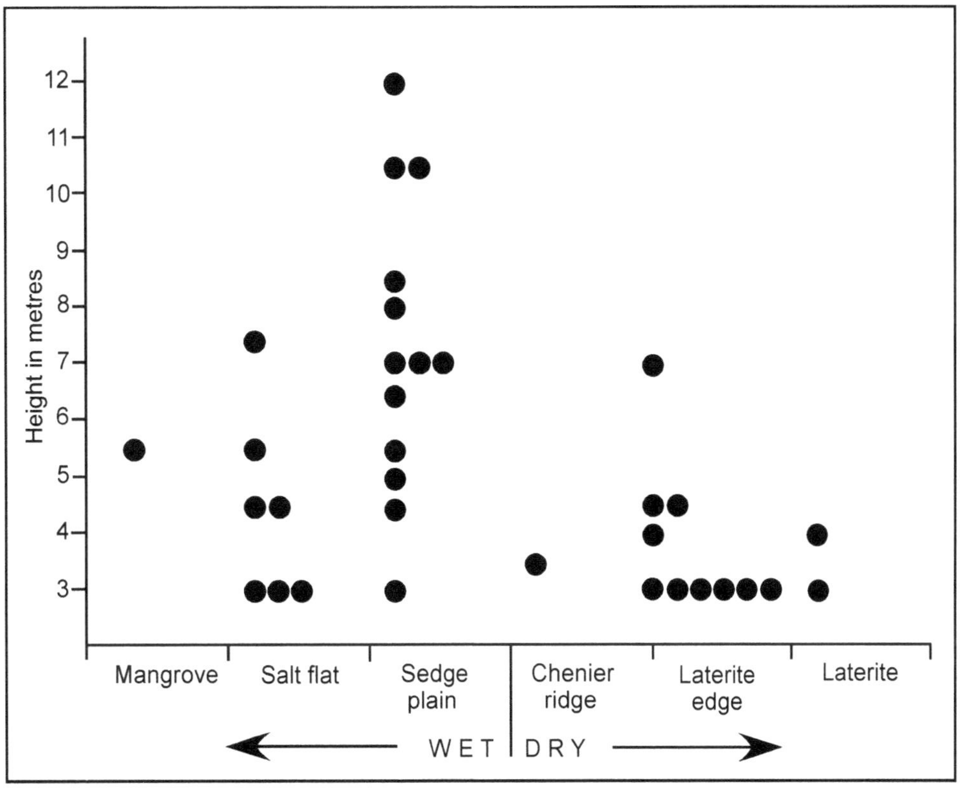

Fig. 11.4. Distribution of Weipa mounds taller than 3 m in relation to different types of substrate and microenvironment. The latter categories correspond to those used in Figure 11.3. Each circle represents a single mound. Note that nearly 21 mounds are on wet surfaces compared to 13 on dry surfaces, while 13 mounds that are 5 m or taller are on wet surfaces compared to only one on a dry surface

consistently associated with open sedge plains (Figure 11.4). There is also some support in the available radiocarbon dates for hypothesis 3, although insufficient excavation or sampling for dates has so far been carried out to enable a fuller examination of inter-site differences in composition and chronology.

The evidence suggests strong support for the hypothesis of self-selection and continued upward growth of shell deposits in a seasonally waterlogged environment (hypothesis 1), at least in the early stages of their growth, but it does not rule out additional factors conducive to mound formation. Once mounds began to form, it seems likely that they would have been recognised as visible features of the landscape, acquired significance in relation to the beliefs and perceptions of the people who used them, and perhaps attracted continued and conscious mounding of the shells in the same place to enhance their symbolic significance in relation to ritual practices. Thus waterlogged ground may be a powerful reinforcement to mounding, but not a sufficient condition by itself. Conversely, waterlogged ground need not be a necessary condition for mounding either. Shell deposition on dry ground tends to create mounds built up of repeated episodes of occupation and shell deposition, even on dry surfaces, though the mounds never achieve the height or volume of sites accumulated on low-lying and seasonally waterlogged surfaces. It seems possible that pre-existing shell deposits form a natural attractor to subsequent re-use, at least up to a point, because of intrinsic advantages, which may include the presence of useful materials abandoned on previous visits, and an attachment by specific families or groups of people to specific places resulting from social tradition and memory. Further investigations of the Weipa mounds themselves may throw further light on these hypotheses, and there is a need for renewed investigations, particularly to expand the sample of radiocarbon dates, which are crucial to discriminating between some of the alternative hypotheses proposed earlier. Another line of investigation would be to examine other parts of the world where massive mounded shell middens have been recorded. Some of these regions have a similar tropical climate to northern Australia, and similar conditions of extensive low-lying waterlogged ground adjacent to productive marine bays and river estuaries, notably in the case of some of the Brazilian shell mounds (Gaspar 1998). But equally, there is considerable evidence, both in Brazil and elsewhere, for a tradition of human burial in shell mounds, for which at present we have no evidence in the case of the large Australian shell mounds.

One of the features of the Weipa shell mounds that needs explanation is their relatively short-lived time span. The present sea level was established in this region about 7000 years ago, and it is tempting to suppose that the late appearance of shell middens is due to time lags in population growth or the development of social practices associated with aggregation and intensification of shellfood consumption (cf. Beaton 1985). Similar shell mounds elsewhere in the tropical north of Australia appear somewhat earlier, but at different dates in different regions, the earliest dates being as early as 4000 BP (O'Connor 1999). It seems likely that this differential pattern has more to do with changes in the palaeogeography and palaeoecology of shorelines than with variable human responses. The shorelines and river estuaries of this tropical environment are highly dynamic in geological and ecological terms. It is quite likely that there was a time lag in some locations, once sea level stabilised, before sufficient sediment accumulated to create the extensive intertidal mudflats that support large beds of *Andara*. Continued accumulation of sediment could have resulted in yet further changes, for example progradation of the shoreline, changes in the sediment composition of the intertidal zone, and the development of thick barriers of mangrove vegetation, all of which could have affected the supply of molluscs, either directly by altering their habitat, or indirectly by changing the ease with which they could be accessed by shellgatherers (Chappell 1982, Chappell & Grindrod 1984, Hiscock 1999).

The above discussion should make it clear that even during a period of relatively stable sea level, a number of factors, both ecological and cultural, can alter patterns of shellfood consumption and shell disposal and hence the surviving pattern of shell midden deposits visible in the archaeological record. What would happen during a period of sea-level change, for example the relatively rapid rise of sea level during the period of deglaciation, or a period of marine regression at the onset of glacial conditions? Here we have to take account not only of the potential ecological disruptions of sea-level change and their likely effects on the availability or abundance of the mollusc supply. We also have to consider the displacement effects of sea level change on the location of shell beds and the optimum places for preparation and consumption of shellfood. During such periods of sea level change, especially on coastlines with a shallow topography, it seems quite possible that the progressive lateral shift of the shoreline and its associated molluscan resources would have been so rapid that, regardless of other factors, there was never an opportunity for a sufficient quantity of shells to accumulate in the same location to create an archaeologically visible shell deposit, let alone the more impressive mounds known from areas such as Weipa (cf. Fischer 1995b). The dynamics of shell midden accumulation and use during earlier periods of sea level change and especially during periods when sea level was lower than the present is one of our biggest areas of ignorance. It is in this respect, perhaps more than in any other, that we need new and targeted studies, the exploration of new territory, and the development of appropriate techniques of underwater investigation.

THE FARASAN ISLANDS, RED SEA BASIN

The challenge of establishing what sort of archaeology is associated with submerged shorelines formed at periods of lowered sea level is currently being met by a variety of methods. The Danes have pioneered methods of underwater excavation in the shallow sheltered waters of Denmark, and have demonstrated both the richness of organic preservation on submerged sites and the large amount of material, with some 3000 sites currently recorded (Fischer 2004, Skaarup & Gron 2004). Similar work has been carried out elsewhere (Faught 2004, Fischer 2005a, Galili *et al.* 1993, Iba *et al.* 1999, Momber 2004). These underwater excavations are largely confined to relatively shallow water at depths of 10 m, accessible to scuba divers, and thus to relatively recent periods associated with the final eustatic rise of sea level in the early Holocene. Exploration of more deeply submerged coastal environments associated with the earlier stages of the glacial cycle is far more technically challenging for diving work and has so far been limited to remote sensing and coring methods aimed at reconstructing features of the ancient terrestrial landscape, with limited recovery of archaeological material (Fedje & Josenhans 2000, Fitch *et al.* 2005). In a new project in the Farasan Islands in the southern Red Sea coastal waters of Saudi Arabia (Figure 11.5), we have been experimenting with mixed gas diving techniques, which allow divers to reach safely much deeper levels down to about 90 m below modern sea level. This work is part of a larger project on the prehistoric archaeology of the region, which is being carried forward by a strategy of integrated archaeological investigation on land and underwater (Bailey *et al.* 2007).

The Farasan Islands are composed of cemented coral platforms that have been uplifted and deformed by salt tectonics. Today they are some 40 km distant from the mainland, but would have been accessible by dry land during the Palaeolithic era at sea levels lower than about –20 m. We have chosen this area for more detailed work for two principal reasons. First, there is a rich and well-preserved on-shore record of postglacial settlement, including many hundreds of shell middens, which can provide insight into the nature of the coastal archaeology associated with the present-day coastline, and a guide to what to look for underwater. Secondly, the Farasan Islands also have deeply embayed shorelines and a complex underwater topography with variable depths of water offshore with quite deep water in the near vicinity, and this offers optimum conditions, relatively sheltered from marine disturbance, for the preservation and discovery of submerged material.

The Holocene shell mounds are located on the present-day shoreline, which usually comprises a coral platform that has been undercut by the chemical solution of the seawater. These undercut coral platforms form a very distinctive feature of the Holocene shoreline, and the first stage of our underwater programme has been devoted to

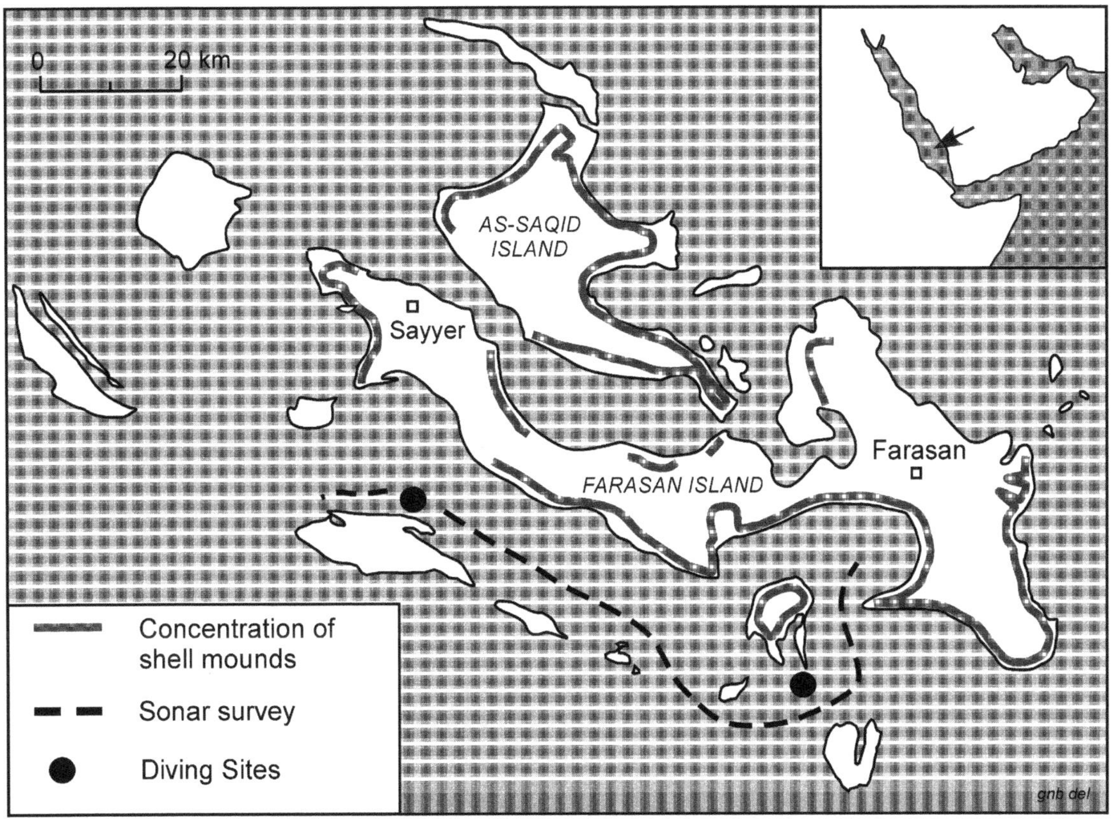

Fig. 11.5. Location map of the Farasan Islands, Saudi Arabia

discovering whether similar features can be found underwater, representing older shorelines formed at lower periods of sea level. The diving programme involves the use of a six-strong diving team trained in the use of nitrox (oxygen and nitrogen) and trimix (oxygen, nitrogen and helium), which enable divers to reach depths safely that would otherwise be inaccessible, the use of a large oil-support vessel (supplied by Saudi Aramco) as a mobile offshore basis, and smaller boats to provide access to diving sites. The diving work has been combined with remote sensing surveys to help in the identification of underwater features, and has been concentrated in near-shore locations with relatively steep drop-offs with good potential for exposure of underwater wave-cut coral platforms similar to those visible onshore and representing earlier shorelines.

The work is still at a very preliminary stage. Our first exploratory surveys have demonstrated the existence of deeply undercut coral platforms representing palaeoshorelines at a variety of depths ranging from 6 m below the present sea level to 60 m. We have also demonstrated the effectiveness of deep diving techniques in tracking these features over some distance, measuring, mapping and photographing them, and collecting samples for dating and other analyses. It is too early for us to say anything definitive about the existence of any associated archaeology, but we expect to continue this work in future years with a more extensive underwater programme combining diving, photography and remote sensing.

CONCLUSION

Shell middens and shell mounds have naturally attracted archaeological attention as indicators of prehistoric coastal economies because of their high visibility and their relative durability and good conditions of preservation in relation to many other sorts of archaeological remains. However, the above discussion leads to the conclusion that the distribution of shell middens and the volume of shells contained within them represent a complex intersection of geological, environmental and cultural variables that affect the visibility and survival of shell deposits independently of variations in subsistence practices or human dependency on marine resources. Absence of shell middens is not in itself evidence for absence of coastal settlement along former shorelines let alone a lack of interest in marine resources more generally. Variable practices of shellgathering and shell disposal may enhance or suppress the visibility of the surviving deposits independently of the quantities of molluscs actually consumed. Moreover, in coastal regions such as Denmark, where favourable conditions of preservation and a long tradition of archaeological investigation have produced a rich and detailed record of Mesolithic coastal settlement and one of the best known regional records of shell mounds comprising hundreds of sites, it is clear that there are a great many coastal sites that lack any shell remains at all (Andersen 2000). This reflects the fact that the most abundant supplies of molluscs in that region are quite localised in their distribution, and that even on productive shorelines they remained only one of many types of marine and terrestrial resources contributing to the overall economy. Even greater uncertainties attach to the absence or rarity of marine shells discarded as food remains in archaeological sites that pre-date the mid-Holocene, because of the likelihood that earlier evidence has been lost because of marine inundation. All of this means that writing a social and economic history of human interest in coastlines and marine resources is going to be far more complicated than we might once have expected. But at the same time, it also means that that history most probably has a much greater time depth and conceals a much more detailed record of variability in human practices than was once assumed to be the case. The technical challenges posed by this view of coastal prehistory and the new horizons that it opens up are likely to provide a rich source of material and ideas to stimulate future investigations.

References

ANDERSEN, S.H. (2000) Kjøkkenmøddinger (shell middens) in Denmark: a survey. Proceedings of the Prehistoric Society 66, p. 361-84.

BAILEY, G.N. (1977) Shell mounds, shell middens and raised beaches in the Cape York Peninsula. Mankind 11, p. 132–43.

BAILEY, G.N. (1993) Shell mounds in 1972 and 1992: reflections on recent controversies at Ballina and Weipa. Australian Archaeology 37, p. 1–18.

BAILEY, G.N. (1994) The Weipa shell mounds: natural or cultural? In Sullivan, M.; Brockwell, S.; Webb, A., eds. Archaeology in the North: Proceedings of the 1993 Australian Archaeological Association Conference. Darwin: North Australian Research Unit (ANU), p. 107–29.

BAILEY, G.N. (2004) The wider significance of submerged archaeological sites and their relevance to world prehistory. In: Flemming N.C., ed. Submarine prehistoric archaeology of the North Sea: research priorities and collaboration with industry. London: CBA Research Report 141, p. 3–10.

BAILEY, G.N. (2007) Time perspectives, palimpsests and the archaeology of time. Journal of Anthropological Archaeology, 26, p. 198-223.

BAILEY, G.N.; CRAIGHEAD, A. (2003) Late Pleistocene and early Holocene coastal palaeoeconomies: a reconsideration of the molluscan evidence from Northern Spain. Geoarchaeology: an International Journal 18: 2, p. 175–204.

BAILEY, G.N.; MILNER, N.J. (2002) Coastal hunters and gatherers and social evolution: marginal or central? Before Farming: the Archaeology of Old World Hunter-Gatherers 3–4: 1, p. 1–15.

BAILEY, G.N.; AL-SHAREKH, A.; FLEMMING, N.; LAMBECK, K.; MOMBER, G.; SINCLAIR, A.; VITA-FINZI, C. (2007) Coastal prehistory in the southern Red Sea Basin: underwater archaeology and the Farasan islands. Proceedings of the Seminar for Arabian Studies 37, p. 1-16.

BEATON, J.M. (1985) Evidence for a coastal occupation time-lag at Princess Charlotte Bay (North Queensland) and implications for coastal colonization and population growth theories for Aboriginal Australia. Archaeology in Oceania 20, p. 1–20.

CHAPPELL, J.; SHACKLETON, N.J. (1986) Oxygen isotopes and sea level. Nature 324, p. 137–40.

CRIBB, R.L. (1986) Archaeology in the Aurukun Region; a report on work carried out in 1985. Queensland Archaeological Research 3, p. 133–58.

ERLANDSON, J.M. (2001) The archaeology of aquatic adaptations: paradigms for a new millennium. Journal of Archaeological Research 9, p. 287–350.

FAUGHT, M.K. (2004) The underwater archaeology of paleolandscapes, Apalachee Bay, Florida. American Antiquity 69: 2, p. 275–89.

FEDJE, D.W.; JOSENHANS, H. (2000) Drowned forests and archaeology on the continental shelf of British Columbia, Canada. Geology 28: 2, p. 99–102.

FISCHER, A., ed. (1995a) Man and sea in the Mesolithic: coastal settlement above and below present sea level. Oxford: Oxbow.

FISCHER, A. (1995b) An entrance to the Mesolithic world below the ocean: status of ten years' work on the Danish sea floor. In: Fischer, A., ed. Man and sea in the Mesolithic: coastal settlement above and below present sea level. Oxford: Oxbow, p. 371–84.

FISCHER, A. (2004) Submerged Stone Age — Danish examples and North Sea potential. In: Flemming, N.C., ed. Submarine prehistoric archaeology of the North Sea: research priorities and collaboration with industry. London: CBA Research Report 141, p. 23–36.

FITCH, S.; THOMSON, K.; GAFFNEY, V. (2005) Late Pleistocene and Holocene depositional systems and the palaeogeography of the Dogger Bank, North Sea. Quaternary Research 64, p. 185–96.

GALILI, E.; WEINSTEIN-EVRON, M.; HERSHKOVITZ, I.; GOPHER, A.; KIZLEV, M.; LERNAU, O.; KOLSKA-HORWITZ, L.; LERNAU, H. (1993) Atlit-Yam: a prehistoric site on the sea floor off the Israeli coast. Journal of Field Archaeology 20, p. 133–57.

GASPAR, M.D. (1998) Considerations of the sambaquis of the Brazilian coast. Antiquity 72, p. 592–615.

HISCOCK, P. (1999) Holocene coastal occupation of western Arnhem Land. In: Hall, J.; Mcniven, I.J., eds. Australian coastal archaeology. Department of Archaeology and Natural History, Research School of Pacific and Asian Studies, the Australian National University, Research Papers in Archaeology and Natural History 31. Canberra: Australian National University, p. 91–103.

IBA, I.; MATSUI, A.; NAKAJIMA, T. (1999) The Awazu shell midden: a submerged site at Lake Biwa. In: Kawanabe, H.; Coulter, G.W.; Roosevelt, A.C., eds. Ancient lakes: their cultural and biological diversity. Belgium: Kenobi, p. 135–45.

MEEHAN, B. (1982) Shell bed to shell midden. Canberra: Australian Institute of Aboriginal Studies.

MOMBER, G. (2004) The inundated landscapes of the western Solent. In: Flemming, N.C., ed. Submarine prehistoric archaeology of the North Sea: research priorities and collaboration with industry. London: CBA Research Report 141. p. 37–42.

MORRISON, M. (2003) Old boundaries and new horizons: the Weipa shell mounds reconsidered. Archaeology in Oceania 38, p. 1–8.

O'CONNOR, S. (1999) A diversity of coastal economies: shell middens in the Kimberley region in the Holocene. In: Hall, J.; Mcniven, I.J., eds. Australian coastal archaeology. Department of Archaeology and Natural History, Research School of Pacific and Asian Studies, the Australian National University, Research Papers in Archaeology and Natural History 31. Canberra: Australian National University, p. 37–50.

SHACKLETON, N.J. (1987) Oxygen isotopes, ice volume and sea level. Quaternary Science Reviews 6, p. 183–90.

SIDDALL, M.; ROHLING, E.J.; ALMOGI-LABIN, A.; HEMLEBEN, C.H.; MEISCHNER, D.; SCHMELZER, I.; SMEED, D.A. (2003) Sea-level fluctuations during the last glacial cycle. Nature 423, p. 853–8.

SKAARUP, J.; GRØN, O. (2004) Møllegabet II: a submerged Mesolithic settlement in southern Denmark. British Archaeological Reports International Series 1328. Oxford: BAR Publishing.

VAN ANDEL, T. (1989) Late Quaternary sea-level changes and archaeology. Antiquity 63, p. 733–45.

DANISH STONE AGE SHELL MIDDENS AND APPLIED GEOARCHAEOLOGICAL AND BIOARCHAEOLOGICAL METHODS

Nina Helt NIELSEN

Department of Prehistoric Archaeology,University of Aarhus, Institute of Anthropology, Archaeology and Linguistics, Moesgaard, 8270 Højbjerg, Denmark, Email: farknn@hum.au.dk

***Abstract**: This paper gives an introduction to Danish Stone Age shell middens, which appear as oblong shell heaps along the prehistoric coast line. Food remains, artefacts and features suggest that daily activities took place at most Mesolithic middens, while the Neolithic middens may have had a more specialised function. Selected examples of how geoarchaeological and bioarchaeological analyses can contribute to the understanding of shell middens and the surrounding environment are presented e.g. how the composition of shell middens can be determined and interpreted, how land snail analysis can help identifying hiatuses in shell middens, and how the marine environment can be reconstructed.*

***Keywords**: Shell middens, Stone Age, Denmark, geoarchaeology, bioarchaeology*

***Résumé**: Cet article donne une introduction à la compréhension des amas d'ustensiles de cuisine danoise de l'âge de la pierre qui se montrent comme des monceaux de coquilles oblongues, le long de la ligne de la côte préhistorique. Des restes de nourritures, des objets et des traits distinctifs indiquent que des activités journalières ont eu lieu sur le site où la plupart des objets de cuisine mésolithiques ont éte retrouvés, tandis que les amas d'ustensiles de cuisine néolithiques ont probablement eu une fonction plus spécialisée. L'article donne des exemples sélectionnés qui expliquent comment les analyses géoarchéologiques et bioarchéologiques peuvent contribuer à une compréhension des amas d'ustensiles de cuisine et de l'environnement. Par exemple, il nous montre comment il est possible de déterminer et d'interpréter la composition des monceaux de coquilles. Il vérifie aussi comment des analyses quant aux escargots qui vivent sur terre peuvent contribuer à l'identification d'hiatus dans les monceaux de coquilles et à la construction du milieu marin.*

***Mots-clefs**: amas d'ustensiles de cuisine, l'âge de pierre, Danemark, géoarchéologie, bioarchéologie*

GEOGRAPHICAL AND TOPOGRAPHICAL SETTING

Denmark is situated in Northern Europe as a kind of barrier between the North Sea and the Baltic Sea (Figure 12.1). Compared to the size of the country (c. 43.000 km^2) Denmark has a very long coastline of 7.300 km and consists of 406 islands. Thus, the coast and the sea, which everywhere is quite shallow, have always been of great importance to the inhabitants of Denmark. This is particularly true for the part of the Stone Age following the global rise in sea level in the early part of the Atlantic (e.g. Christensen 1995).

Due to eustatic and isostatic processes the outline of Denmark during the Atlantic period was somewhat different from what it is today but at that time Denmark also consisted of numerous islands, shallow inlets etc. (Figure 12.2). The marine environment was extremely rich and the marine resources were fully exploited by the people living here in the Stone Age. This is demonstrated from the several finds of fish hooks, leisters, harpoons and fish traps, ^{13}C-content of bones from humans and dogs as well as from the direct evidence of food remains such as charred food crusts, bones from fish and sea mammals in addition to marine molluscs. In some places the gathering of marine molluscs was so intensive or was carried out on the same location for hundreds or thousands of years that it resulted in the accumulation of large shell middens.

The Danish Stone Age shell middens are primarily found in the Limfjord area and on the coasts facing the Kattegat (Figure 12.2) as this is where the marine conditions such as the salinity level have been most suitable for e.g. oysters, *Ostrea edulis*, which is one of the primary species found in the middens.

DANISH SHELL MIDDENS

Dating

The oldest shell middens date back to around 5600 cal BC i.e. the last part of the Kongemose Culture. However, it is important to note that exploitation of marine resources probably also took place in earlier periods, but that the evidence of this practise today is hidden far below the sea. Most Danish middens date from the Late Mesolithic Ertebølle culture (5400-3950 cal BC) – particularly around 4600-4000 BC – or from the Neolithic Funnel Beaker Culture (3950-2800 cal BC) predominantly from the earliest part, 3950-3450 cal BC (Andersen 2004, 2008). Several middens contain material from both periods and do therefore represent a very good source for investigating the transition from the Mesolithic to the Neolithic, e.g. Bjørnsholm and Norsminde (Andersen 1991, 1993). Shell middens are also found in low numbers during the Pitted Ware and Single Grave Culture, 2800-2350 cal BC, and in higher numbers later during the Iron Age (Andersen 2008).

Shape and size

Danish shell middens do not form large mounds but can often be seen as oblong heaps stretching along the old

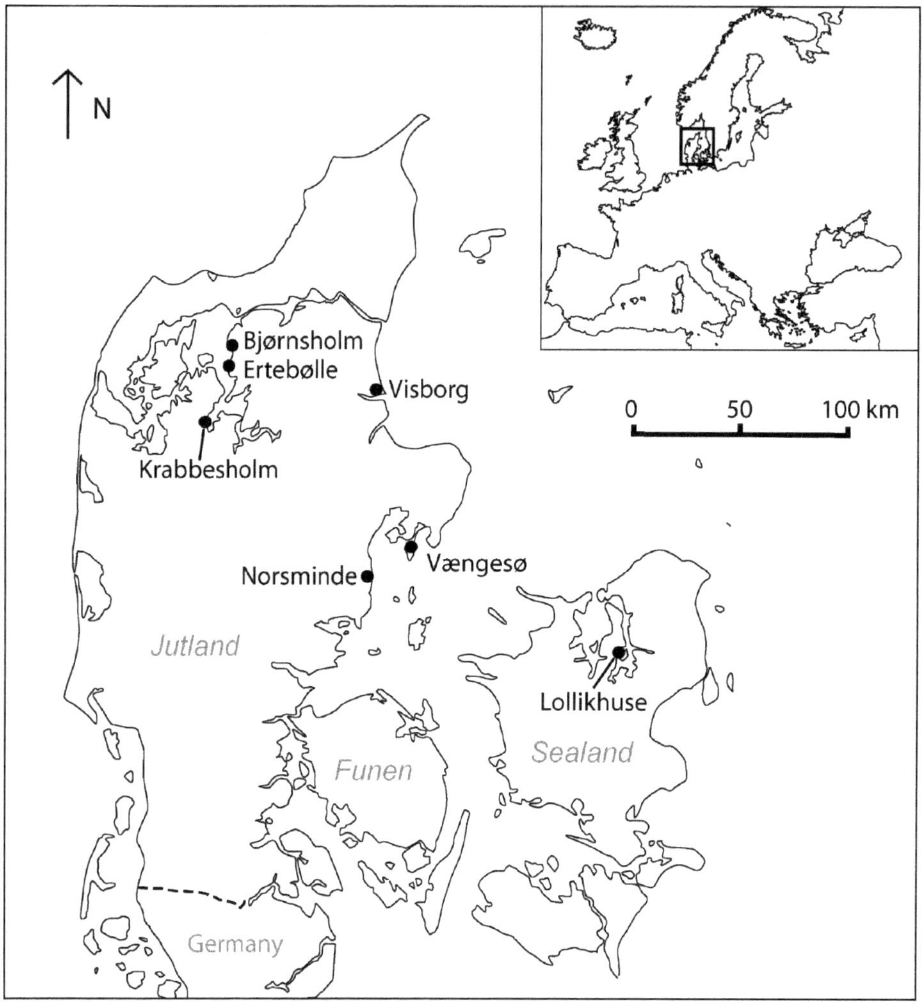

Fig. 12.1. Map of Denmark showing the location of some of the most important shell middens, which also are mentioned in the text

coastline. The middens can be up to 1.9 m thick as the Ertebølle midden and stretch along the coast for up to 700 m like the Visborg shell midden (Madsen *et al.* 1900, 20; Andersen 2000a). The largest and thickest middens often cover a time span of hundreds or thousands of years and are situated in the Limfjord area and in eastern Jutland, e.g. Ertebølle, Bjørnsholm, Krabbesholm and Norsminde (Andersen & Johansen 1987; Andersen 1991, 1993, 2005). However, the Danish middens are very variable in size and can also be very local in distribution or consist of just one thin shell layer – it very much depends on the number of visits that have been paid to the site and the duration of the settlement. Some of the thinner middens are, however, the result of later disturbances such as modern cultivation. Thus, for instance, at the Vængesø III midden only the shell layers lying in a shallow depression where the plough could not reach were preserved.

Composition and interpretation of the middens

Danish shell middens are usually defined as being deposits dominated by marine molluscs and forming a continuous horizon with a cubic content of at least 10 m^3 (Andersen 2000b, 362). Generally the middens consist of at least 70-80 % of shells; the most common molluscs, which are also the species assumed to have been exploited as food sources, being oysters (*Ostrea edulis*), mussels (*Mytilus edulis*), cockles (*Cerastoderma edule*) and periwinkle (*Littorina littorea*). In addition species such as carpet shell (*Venerupis decussatus*, *V. pullastra* and *V. aureus*), netted dog whelk (*Nassarius reticulatus*), needle whelk, (*Bittium reticulatum*), peppery furrow shell (*Scrobicularia plana*), Baltic telling (*Macoma balthica*) etc. occur more or less incidentally in the deposits (e.g. Nielsen 2008).

In most Mesolithic middens flint debris, flint, bone and antler artefacts, ceramics (after c. 4600 cal BC), potboilers, charcoal and animal bones from terrestrial mammals, fish, birds and sea mammals are very common. Hearths of different types are also frequent and often a clear stratigraphy can be observed as the hearths repeatedly have been placed on the same location. At some sites hearths have furthermore been found evenly spaced along the midden and these distribution patterns clearly indicate that a general organisation of the

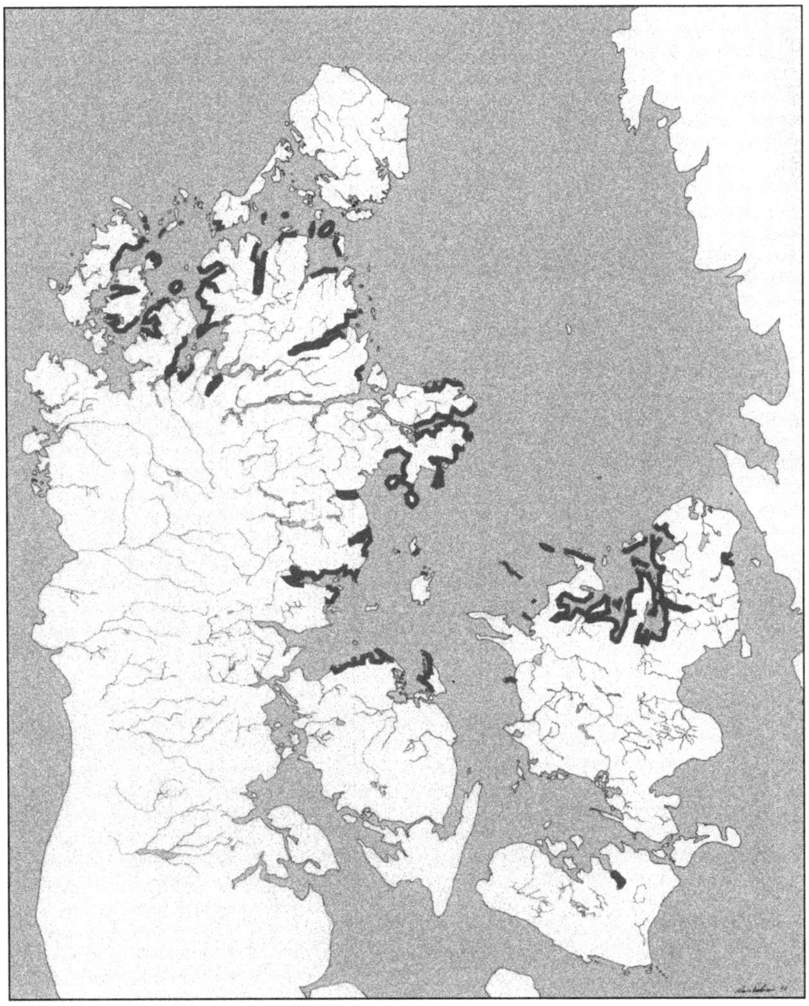

Fig. 12.2. Map of Denmark during the Atlantic period showing the coastlines along which Stone Age shell middens occur (shaded)

settlement persisted during several occupation episodes (Andersen & Johansen 1987; Andersen 2004, 404). Other features such as pits, post holes etc. are less common and evidence of dwellings have never been found in the actual shell layers or in the areas behind the middens. However, possible hut structures with sunken floors have been found at Lollikhuse, Åle and probably at Vængesø III (Sørensen 1995, 24; Andersen 2004, 394). These huts belong to the so-called pre-midden layers found below many shell middens and they are therefore seen as representing habitations prior to the gathering of shell fish (Andersen 2004, 397). However, it seems plausible that some of these huts and pre-midden layers may have been in use when the shell gathering was initiated. At that time the shells would have been deposited somewhere outside the huts, and not until a substantial amount of shells had been gathered would the pre-midden layers and possible huts have been covered by shell layers. Furthermore, the fact that the pre-midden layers rarely form continuous horizons below shell middens but usually are only present in smaller areas (Andersen 2004, 397) supports this interpretation as the areas without the pre-midden layers could represent the areas first covered by shells. However, this interpretation needs to be confirmed by more ^{14}C-datings and targeted research.

Seasonality studies of animal bones and marine molluscs show that the large shell middens have been used during several seasons (Madsen *et. al.* 1900, 175; Bratlund 1993; Milner 2002, Bailey, G.N. & N. Milner *in press*). In addition, the middens often contain a wide variety of artefacts, which indicate that a broad range of everyday activities took place on the site such as cooking, tool making, and activities involving the use of different tools. The large middens are therefore generally interpreted as actual settlements or base camps (Andersen 2000b, 375). On the other hand, some of the smaller middens contain only a few artefacts, a specialised content of animal bones and/or have been used during only one season of the year why these are considered as specialised camps e.g. Aggersund (Møhl 1979). One problem, however, is that the large shell middens are made up of shell deposits accumulated during several occupation episodes and it is therefore hard to tell whether the site has been in use during the whole year or rather several times during one or more years. It will thus be necessary to separate out the archaeological material associated with the individual depositional events in order to establish which activities that took place during the different occupation episodes. It is plausible that the large middens during some visits functioned as a specialised camp where people came to the site in order to fish or gather oysters, while the site at other times functioned as a kind of base camp. That the function of the site can change over time is perhaps obvious, but nevertheless important to bear in mind, just as these circumstances highlights the importance of detailed stratigraphical analyses of the shell middens such as those mentioned below. Furthermore, it is important to note that shell middens were only a part of the wider Mesolithic settlement system consisting of several types of sites of which the non-shell bearing sites were the most common (e.g. Andersen 1995).

Neolithic middens often contain many pot boilers and animal bones but only a few artefacts compared to the Mesolithic middens. Although thick "clean" shell layers occur in Neolithic middens the stratigraphy often consist of thin alternating layers of shells and layers of ash and charcoal. Thus it seems that the function of the shell middens changed somewhat in the Neolithic (e.g. Andersen 2008). Instead of representing actual living floors where a wide variety of activities were carried out, the Neolithic shell middens primary seem to contain refuse connected with the processing and preservation of marine resources and terrestrial mammals, which e.g. is indicated by the many potboilers. Despite these possible changes the presence of Neolithic shell middens clearly indicate a continued exploitation of marine resources into at least the early part of the Neolithic – a subject which, however, have been widely debated on the basis of ^{13}C-analyses during the last years (Milner *et al.* 2004, 2006; Hedges 2004; Barbarena & Borrero 2005, Richards & Schulting 2006).

Behind the midden of Norsminde and Bjørnsholm culture layers dating from the Neolithic were found at a distance of 50 m and 20-40 m respectively from the shell midden on the seashore, which may represent the remains of the actual settlements associated with the shell middens (Andersen 1991, 17; 1993, 65). However, the Early Neolithic settlement system and the role of the shell middens within this system are still very sparsely elucidated. Settlements with substantial evidence of structures etc. are very rare and usually pits are the only remains that are found. The few Early Neolithic settlement sites thus appear in great contrast to the contemporary shell middens which, among other things due to their high visibility in the landscape, are found in a high number.

GEOARCHAEOLOGICAL AND BIOARCHAEOLOGICAL INVESTIGATIONS

The investigation of Danish shell middens began more than 150 years ago and was from the very beginning carried out in collaboration between archaeologist and experts in the natural sciences such as geology, botany, zoology and malacology formalized under the so called kitchen midden commissions (Kristiansen 2002). This interdisciplinary tradition within the shell midden research in Denmark has continued in the investigations of the 20th and 21st century.

Composition of the shell middens

Already in the 19th century the excavators realized that the middens consist of several smaller individual shell heaps (Figure 12.3; Madsen *et al.* 1900, 22, 28). During the more recent excavations it has been established that these heaps usually have a diameter of 2-7 m (Andersen 2004, 406), although smaller local concentrations also occur. The heaps can usually be identified on the basis of the composition, colour or compactness of the shell layer, which differs from the superimposing and underlying layers.

The composition of most of the large and systematically excavated middens has been examined in detail by analyzing a column sample from the midden (Nielsen *in* press A, Bailey & Milner *in press,* Petersen 1987). The samples of e.g. 50x50 cm can be taken while excavating and documenting all finds in the column, which reduces the interference with the overall excavation to a minimum. After taking the samples the individual spits can be studied in detail in the laboratory and e.g. the composition of the shell layers can be established by sieving the sample and then sort, count and weigh the different components.

The shell layers sometimes consist of primarily one shell species, but at other times the layers contain a couple of species such as mussels and oysters combined. In some middens naturally deposited shell layers occur as well, which are very important to distinguish from the anthropogenic deposits. This can often be done by examining the age distribution in the molluscan assemblage, as the anthropogenic layers usually are dominated by the older and larger individuals while the entire age spectra tend to occur in the natural deposits. The content of minerogenic material varies from midden to midden. Thus, in the Ertebølle midden clay and silt make up more than 50 % in some layers (Petersen 1987), while these particles only constitute a few percentages in other middens. The number of pebbles also varies and can be the result of natural deposition, deliberate collection by humans or the incidental transport by humans in connection with e.g. gathering of molluscs or seaweed.

One of the most recent investigations of a Danish shell midden is that of the Mesolithic–Neolithic shell midden, Krabbesholm II situated in the innermost part of the Limfjord (Nielsen 2004; 2008). The analysis of a column sample from this site yielded a detailed picture of the changing composition of the shell layers through time (Figure 12.4) and it was even possible to identify an

Fig. 12.3. Section of the Ertebølle shell midden during the excavations in the 1890s. Notice the easily distinguishable layers dominated by oysters and cockles respectively

additional shell layer in the brown earth soil, which had not been recognized during excavation due to the high degree of bioturbation.

Results from the analysis of a column sample will naturally always show the average composition within the individual spits making up the column. However, when taking the column sample at Krabbesholm II it was sometimes possible to discern a very localised and only a few mm thick layer of e.g. mussels within an oyster layer. However these layers were too thin to be sampled individually and do therefore not show up on the diagrams that were produced. These thin layers within the otherwise well defined shell heaps show that the heaps were probably not from one but rather from several gathering events – an interpretation which possibly could be supported by seasonality studies of the marine molluscs (Milner 2001).

Identification of hiatuses

To identify the individual episodes of accumulation as well as hiatuses in the deposits is very important in order to understand the formation and the use of shell middens. ^{14}C-datings have contributed greatly to the overall dating of the middens in Denmark. However, the margin of errors associated with this dating method is far too great to produce a fine chronology and reconstruct the accumulation history of the middens.

Danish middens rarely have layers of windblown sand or other sediments separating the different episodes of shell accumulation. Until now the identification of hiatuses in the midden accumulation therefore have primarily been based on a high abundance of artefacts and the presence of hearths and ash layers, which is thought to indicate a pause in the accumulation of shells (e.g. Andersen & Johansen 1987, 44; Andersen 1993). The problem with this method, however, is that a hearth and even several flint artefacts in theory can represent the remains of activities taken place on a single afternoon. A further problem is that some middens – particularly the Neolithic middens as mentioned above – contain very few artefacts, which makes it impossible to identify breaks in the accumulation of shells based on these. However, a new method may have been found which can help addressing the problem of identifying hiatuses.

During the investigation of Krabbesholm II, the author carried out a study on land snails from the column sample. The main purpose for carrying out the analysis was to reconstruct the terrestrial environment during the occupation, which indeed proved to be very successful (Nielsen 2004, 2007). This is not least due to the exceptional good preservation conditions for land snails in shell middens and the method definitely holds great potential for reconstructing the local environment – in particularly as e.g. pollen analysis cannot be carried out in these very alkaline deposits. However, in addition to reconstructing the environment, it turned out that land snail analysis can be used to identify breaks in the accumulation of shell middens as the abundance of snails in a shell layer to a high extend will depend on the accumulation rate of the deposit. Thus a high accumulation rate will lead to a low abundance of snails, while a low accumulation rate generally will result in a high abundance of land snails.

In the study of Krabbesholm II few snails were generally present in the lower and middle part of the individual

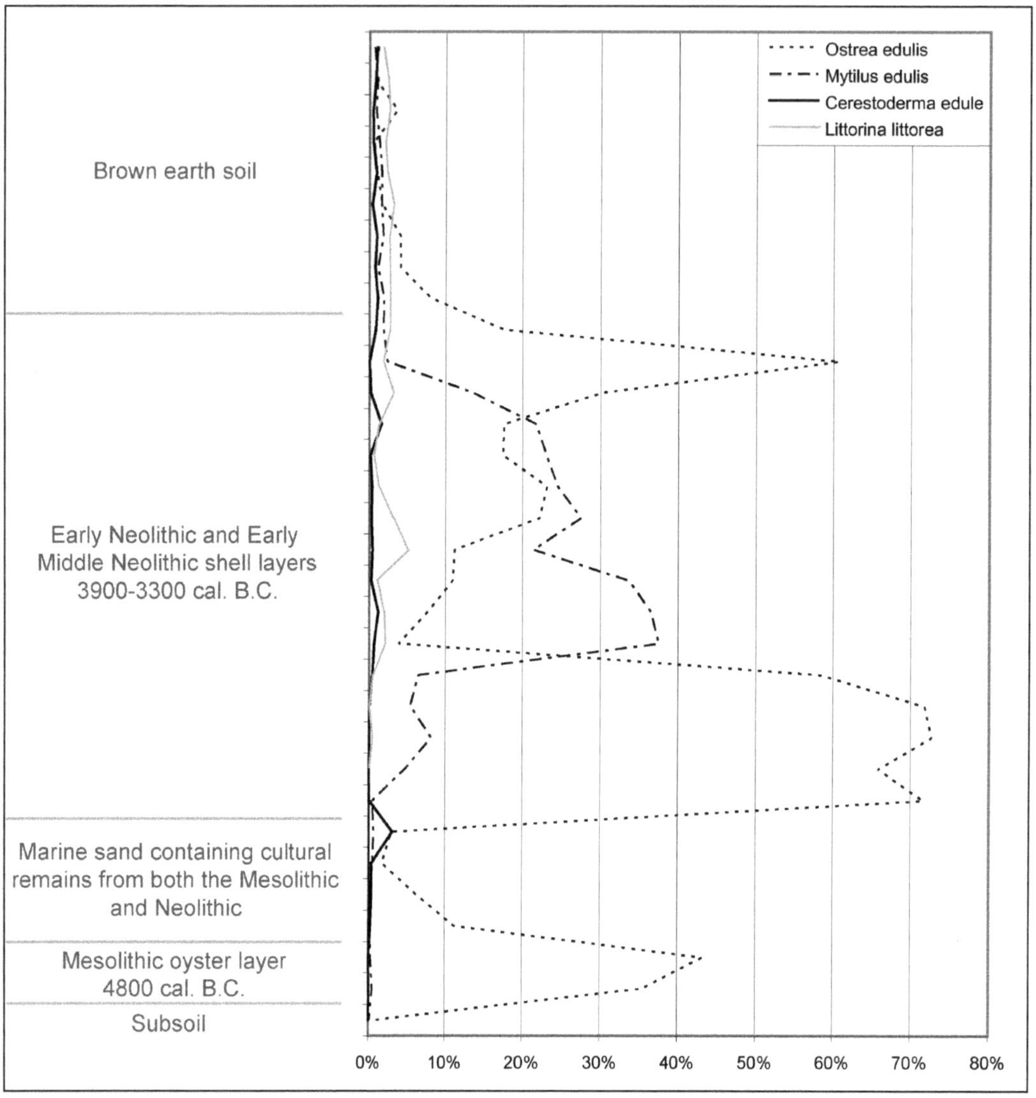

Fig. 12.4. Diagram showing the species composition in the main part of the Krabbesholm II shell midden. The graph is based on data resulting from a detailed examination of a column sample which involved sieving, sorting and weighing of the material. Only shell fragments larger than 6 mm are included in the diagram

shell layers, while the top part of some of the layers was very abundant in snails. This implies that the surface on these layers must have been open for some time in order for the snails to encroach into the area and accumulate in the deposit, which means that it could possibly represent periods where the site was uninhabited or at least periods where accumulation of the midden paused. On the other hand, the low number of snails in the middle part of the shell layers can be explained as being due to a quite rapid accumulation of marine shells within the individual layers. That a high concentration of land snails in one case is associated with a very crushed layer indicating trampling clearly supports this hypothesis.

If land snail analysis is carried out on only one column sample from a midden, as was the case with Krabbesholm II, it is difficult to establish whether the pauses in accumulation represent general breaks in the occupation of the midden or whether accumulation of marine shells continued some metres from where the sample was taken. However, it seems that a new approach, which can help interpret the complex stratigraphy of shell middens have been found and deserves to be further developed and explored.

Minerogenic layers in shell middens

As mentioned above, it is very important to recognize and sort out naturally deposited shell layers in shell middens. However, in shell middens where minerogenic layers are present between shell layers, it is also very important to establish the origin of these as they will reflect the kind of processes that have been going on at the site. At Krabbesholm II geoarchaeological investigations including grain size analysis, soil micromorphology, loss on ignition, phosphate analysis and analysis of foraminifera were carried out in addition to the molluscan studies on two minerogenic layers (Nielsen 2004). Based on these analyses as well as on the field observations it was concluded that the minerogenic layer between the Mesolithic and the Neolithic shell layer probably had a marine origin and therefore indicated a period with a higher sea level. On the other hand, a minerogenic layer dating from the Single Grave Culture (ca. 2800 BC) further up in the midden most likely represented colluvium deposited during a period where the area behind the midden must have been very open, which was also indicated by the land snails (Nielsen 2007). The soil micromorphological investigation also provided insight into the post-depositional processes taking place in the shell midden. Translocated calcium carbonate was present in all the thin sections (which were only taken from minerogenic layers) indicating that the shells are being subject to chemical decomposition. However, as fish bones are present in the minerogenic layer separating the Mesolithic and the Neolithic shell layer the preservation conditions must still be quite good. Bioturbation was also evident in all thin sections, which among other things was observed to result in mechanical breaking down of charcoal. Another interesting micromorphological result derived from the analysis of a hearth. Here it could be observed that fresh as well as decayed wood had been burned and that the hearth had contained a lot of phosphatic material. It was even possible to identify fragments of meat and some kind of plant material, which had been prepared in the hearth (pers. com. M.-A. Courty). These results are naturally not that surprising, but still it is interesting when you can obtain insight into the use of e.g. a hearth at such a detailed level.

Micromorphology is a method almost completely unknown within Danish archaeology and no analysis has yet been made on the actual shell layers as at Tierre del Fuego (Vila *et al.* this volume). However, hopefully in the future it will be possible to make more analyses on the Danish shell middens as this method can provide a detailed knowledge of the formation and composition of the middens, the post-depositional processes that have influenced on the deposits and of specific features in the middens, which is not possible to obtain through the use of the conventional archaeological and sedimentological methods.

Marine environment

Marine conditions determine the marine fauna and do thereby also indirectly affect the anthropogenic formation of shell middens. Reconstruction of the marine conditions in the Stone Age, such as sea level fluctuations and the salinity and temperature of the sea, is therefore another important research area when studying shell middens.

The marine molluscs in shell middens do to some degree reflect the marine environment. Stable oxygen and carbon isotope analyses of *Littorina littorea* specimens from the Ertebølle shell midden thus proved to be suitable for reconstructing the marine environment. The analyses showed that the summer surface temperature of the sea during the Atlantic period was about 22°C i.e. 2-4°C warmer than today and with a salinity of about 31 PSU, which can only be explained by a more pronounced water exchange between the Limfjord and the North Sea than today (Burman & Schmitz 2005). Also the bare presence of molluscs with specific temperature and salinity requirements such as oysters, carpet shells and *Bittium reticulatum* can be useful. Thus on the basis of this it could be demonstrated that even in the innermost part of the Limfjord, where Krabbesholm II is situated, the sea remained saline and warm during at least the Early Neolithic and the early part of the Middle Neolithic (Nielsen 2008). However, one should be aware of the several examples showing that the absence of a particular species in a shell midden does not necessarily imply that the species is not occurring naturally in the sea. For instance factors such as human preferences can influence on the selection of exploited species (Nielsen 2008).

Investigations of natural shell banks from the same period as the shell middens can be very rewarding and produce more reliable information on the naturally occurring marine molluscs. Today hundreds of natural shell banks and marine deposits containing marine molluscs are known – e.g. one just 200-300 m away from the classical Ertebølle midden – but only a few of these have been thoroughly investigated or dated (e.g. Petersen 1987, 79; Rasmussen & Petersen 1980, 51). A large unexploited material is thus waiting to be investigated in Denmark.

Analyses of cores from the old seabed is furthermore an important method for obtaining detailed information on the changing marine environmental conditions e.g. on the basis of microorganisms such as diatoms, but preferably through a multi-proxy approach. For instance at Norsminde Fjord new investigations are currently being undertaken, which are hoped to confirm the hypothesis that the exchange of seawater into the small and narrow inlet was reduced around the transition to the Neolithic (Rasmussen *in press* and the following chapters). This hypothesis is based on a shift in exploitation pattern in the Norsminde shell midden from primarily oysters in the Mesolithic to primarily cockles in the Neolithic, which are thought to have been caused by the less favourable conditions for oysters. However, at other places in Denmark where the access to fresh sea water remained open, oysters continued to be exploited well into the Neolithic e.g. at Visborg and at Krabbesholm as mentioned above (Nielsen 2008).

CONCLUSION

Denmark has a long tradition for investigating shell middens, which has resulted in a detailed knowledge of this phenomenon. The conventional archaeological analyses will always provide the basis for the investigations; however, there is no doubt that geoarchaeological and bioarchaeological analyses can expand the knowledge of the shell middens. Old methods, such as land snail analysis, have been taken up and provided very interesting results; other methods are being refined or developed, e.g. chemical analyses, and open up for new interpretations and research areas. These geoarchaeological and bioarchaeological analyses can provide a detailed insight into various aspects of the prehistory which it is not possible to obtain through conventional archaeological analysis and it is thus very important to encourage the development within these research areas.

The shell middens in Denmark are disappearing very fast and only few middens are classified as protected prehistoric monuments. This increases the importance first of all of working for the protection of more shell middens and secondly of investigating the shell middens with the greatest care and with a clear objective in mind in order to obtain the highest possible outcome of the studies. To use the right methods and strategies is thus also of great importance but these can only be chosen when having a knowledge about the different available methods and their advantages and disadvantages. It is therefore also important that the valuable experiences and results are being shared among researchers and that research projects involve a broad range of experts. This is to some degree happening today, but unfortunately the research environment in e.g. Denmark is extremely closed, which clearly restrains the outcome of the research.

Acknowledgement

I thank Mette Løvschal for the French translation of the abstract.

Bibliography

ANDERSEN, S.H. (1991) – Norsminde. A "Køkkenmødding" with Late Mesolithic and Early Neolithic Occupation. Journal of Danish Archaeology. Odense. 8:1989, p. 13-40.

ANDERSEN, S.H. (1993) – Bjørnsholm. A Stratified Køkkenmødding on the Danish Central Limfjord, North Jutland. Journal of Danish Archaeology. Odense. 10:1991, p. 59-96.

ANDERSEN, S.H. (1995) – Coastal adaption and marine exploitation in Late Mesolithic Denmark – with special emphasis on the Limfjord region. In Fischer, A., ed. – Man and Sea in the Mesolithic Coastal settlement above and below present sea level. Proceedings of the International Symposium Kalundborg, Denmark 1993. Oxford: Oxbow, p. 41-66. (Oxbow Monograph 53).

ANDERSEN, S.H. (2000a) – Fisker og bonde ved Visborg (with English summary). In Hvass, S.; Det Arkæologiske Nævn, eds – Vor Skjulte kulturarv. Arkæologien under overfladen. Til Hendes Majestæt Dronning Margrethe II 16. april 2000. Copenhagen/ Højbjerg: Det Kongelige Nordiske Oldskriftselskab/ Jysk Arkæologisk Selskab, p. 42-43, 208.

ANDERSEN, S.H. (2000b) – "Køkkenmøddinger" (Shell Middens) in Denmark: a survey. Proceeding of the Prehistoric Society. Cambridge. 66, p. 361-384.

ANDERSEN, S.H. (2004) – Danish Shell Middens Reviewed. In Saville, A, ed – Mesolithic Scotland and its Neighbours. The Early Holocene Prehistory of Scotland, its British and Irish Context, and some Northern European Perspectives. Edinburgh: Society of Antiquaries of Scotland, p. 393-411.

ANDERSEN, S.H. (2005) – Køkkenmøddingerne ved Krabbesholm. Ny forskning i stenalderens kystbopladser. Nationalmuseets Arbejdsmark. København. 2005, p. 151-171.

ANDERSEN, S.H. (2008) – Shell middens *("Køkkenmøddinger")* – the Danish Evidence. In Antczak, A.; Cipriani, R., eds – Early Human Impact on Megamolluscs. Oxford: BAR international series 1865, p. 135-156.

ANDERSEN, S.H.; JOHANSEN, E. (1987) – Ertebølle Revisited. Journal of Danish Archaeology 5:1986, p. 31 61.

BAILEY, G.N.; MILNER, N. (*in press*) – The marine molluscs from the Mesolithic and Neolithic deposits of the Norsminde shell midden. In Andersen, S., ed – Stone Age Settlement in the Coastal Fjord of Norsminde, Jutland, Denmark.

BARBARENA, R.; BORRERO, L.A. (2005) – Stable isotopes and faunal bones: Comments on Milner *et al.* (2004). Antiquity. Gloucester. 79:303, p. 191-195.

BRATLUND, B. (1993) – The bone remains of mammals and birds from the Bjørnsholm shell-mound. Journal of Danish Archaeology. Odense. 10:1991, p. 97-104.

BURMAN, J.; SCHMITZ, B. (2005) – Periwinkle (*Littorina littorea*) intrashell $\delta^{18}O$ and $\delta^{13}C$ records from the mid-Holocene Limfjord region, Denmark: a new high-resolution palaeoenvironmental proxy approach. The Holocene. London. 15:4, p. 567-575.

CHRISTENSEN, C. (1995) – The *Littorina* Transgressions in Denmark. In Fischer, A., ed – Man & Sea in the Mesolithic. Coastal settlement above and below present sea level. Oxford: Oxbow, p. 15-22. (Oxbow Monograph 53).

HEDGES, R.E.M. (2004) – Isotopes and red herrings: comments on Milner *et al.* and Lidén *et al.* Antiquity. Gloucester. 78:299, p. 34-37.

KRISTIANSEN, K. (2002) – The Birth of Ecological Archaeology in Denmark: history and research environments 1850-2000. In Fischer, A.; Kristiansen, K., eds – The Neolithisation of Denmark. 150 years of debate. Sheffield: J.R. Collins Publications, p. 11-31.

MADSEN, A.P.; MÜLLER, S.; NEERGAARD, C.; PETERSEN, C.G.J.; ROSTRUP, E.; STEENSTRUP K.J.V.; WINGE, H. (1900) – Affaldsdynger fra Stenalderen I Danmark undersøgte for Nationalmuseet. Copenhagen: Reitzel.

MILNER, N. (2001) – At the cutting edge: using thin sectioning to determine season of death of the European oyster, *Ostrea edulis*. Journal of Archaeological Science. London. 28, p. 861-873.

MILNER, N. (2002) – Incremental growth of the European oyster, Ostrea edulis: seasonality information from Danish kitchenmiddens. Oxford: BAR. (BAR International Series 1057).

MILNER, N.; CRAIG, O.E.; BAILEY, G.N.; PEDERSEN, K.; ANDERSEN, S.H. (2004) – Something fishy in the Neolithic? A re-evaluation of stable isotope analysis of Mesolithic and Neolithic coastal populations. Antiquity. Gloucester. 78:299, p. 9-22.

MILNER, N.; CRAIG, O.E.; BAILEY, G.N.; ANDERSEN, S.H. (2006) – A response to Richards and Schulting. Antiquity. Gloucester. 80:308, p. 456-458.

MØHL, U. (1979) – Aggersund-bopladsen zoologisk belyst. Svanejagt som årsag til bosættelse (with English summary). Kuml – Årbog for Jysk Arkæologisk Selskab. Højbjerg. 1978, p. 57-75.

NIELSEN, N. (2004) – Geoarchaeological investigations at the Danish kitchen midden Krabbesholm: A depositional and environmental history. Unpublished MSc dissertation from The University of Reading.

NIELSEN, N. (2007) – Land snails and shell middens: A new approach in Danish archaeological research. In Milner, N.; Craig O.; Bailey G., eds – *Shell middens in Atlantic Europe*, Oxford: Oxbow books, p. 70-77.

NIELSEN, N. (2008) – Marine molluscs in Danish Stone Age middens; A case study on Krabbesholm II. In Antczak, A.; Cipriani, R., eds – Early Human Impact on Megamolluscs. Oxford: BAR international series 1865, p. 157-167.

PETERSEN, K.S. (1987) – The Ertebølle "*Køkkenmødding*" and the Marine Development of the Limfjord, with Particular Reference to the Molluscan Fauna. Journal of Danish Archaeology. Odense. 5:1986, p. 77-84.

RASMUSSEN, L.A.; PETERSEN, K.S. (1980) – Resultater fra DGU´s genoptagne kvartærgeologiske kortlægning. Dansk geologisk Forenings, Årsskrift for.... København. 1979, p. 47-54.

RASMUSSEN, P. (*in press*) – The terrestrial and marine palaeoenvironment at Norsminde Fjord, East Jutland, c. 6900-1700 cal. yr BC: Study aims, site description, formation of the fjord, sedimentation history and hydrography. In Andersen, S., ed – Stone Age Settlement in the Coastal Fjord of Norsminde, Jutland, Denmark.

RICHARDS, M.P.; SCHULTING, R.J. (2006) – Against the grain? A response to Milner *et al.* (2004). Antiquity. Gloucester. 80:308, p. 444-456.

SØRENSEN, S.A. (1995) – Lollikhuse – a dwelling site under a kitchen midden. Journal of Danish Archaeology. Odense. 11:1992/1993, p. 19-29.

VILA, A.; ESTÉVEZ, J.; PIANA, E.; MADELLA, M.; BARCELÓ, J.A.; ZURRO, D.; CLEMENTE, I.; TERRADAS, X.; VERDÚN, E.; PIQUE, R.; MAMELI, L.; BRIZ, I. (in press) – Microstratigraphy of shell middens of Tierra del Fuego. This Volume.

A OCUPAÇÃO DOS GRUPOS PRÉ-HISTÓRICOS DE PESCADORES-COLETORES EM PAISAGENS INSULARES DA COSTA DO ESTADO DE SÃO PAULO

Manoel M.B. GONZALEZ

Núcleo de Pesquisa e Estudo em Chondrichthyes – Centro Regional de Pesquisas Arqueológicas – NUPEC/CERPA, Rua Ana Pimentel 12, 11055-130, Santos, São Paulo, Brasil. e-mail: gonzalez@nupec.com.br

Sandra Nami AMENOMORI

Procuradoria Geral da República, Ministério Público Federal. SAF Sul, Quadra 4, Conjunto C, Bloco B, sala 302, 70050-900, Brasilia, DF, Brasil. e-mail: sandranami@gmail.com

***Abstract**: The occupation of prehistoric fishing-gatherer groups in insular landscape at the coast of Sao Paulo State, Brazil. The aim of this work is to study aspects of the prehistoric insular landscape, based on "landscape archaeology and human geography theories", with the purpose to verify prehistoric fisher-gatherer occupation in eight archaeological sites at the coast of Sao Paulo state. The analysis of these archaeological sites was focused on the landscape with all its natural and cultural features. This work is based on the thesis that the groups who lived there shared the same landscape perception: a place that could offer shelter and hide. This symbolic meaning was important when they were supposed to choose a settling place. Even when the new area was not surrounded by water, the apparent aspect of insular landscape given by the shape of a mound caused a sensation of safety and isolation among these groups and at the same time there is a meaning of notability in the landscape.*

***Keywords**: Shell middens, fishing-gatherers, simbolism, island, Brazil*

***Resumo**: O principal objetivo deste trabalho foi estudar os aspectos pré-históricos e insulares baseados em "teorias da arqueologia da paisagem e geografia humana", com o propósito de verificar a ocupação pré-histórica dos grupos pescadores-coletores em oito sítios arqueológicos do litoral do Estado de São Paulo. A análise destes sítios arqueológicos foi enfocada na relação entre a paisagem e as características naturais e culturais. Os resultados foram baseados na tese que viviam no litoral possuíam a mesma percepção da paisagem: um lugar que oferecesse abrigo e funcionasse como ponto de observação. Este simbolismo foi muito importante em momentos de escolha e troca de ambientes. Quando a área escolhida não está rodeada por água, podemos perceber que os sambaquis possuem aspectos de ilhas, causando para os grupos de pescadores-coletores a sensação de segurança e isolamento.*

***Mots clés**: Amas coquilliers, pêcheurs-collecteurs, simbolism, île, Brésil*

INTRODUÇÃO

Entre seis mil anos atrás e o primeiro milênio da era cristã, o litoral brasileiro – inclusive o segmento paulista, neste caso representado pela Baixada Santista e litoral norte – foi intensamente povoado no passado por grupos indígenas, bastante adaptados ao meio ambiente e portadores de eficiente tecnologia para a confecção de instrumentos que utilizavam na caça e na pesca. Respeitavam os mortos, que eram enterrados em locais específicos com vários acompanhamentos funerários e muitos indícios levam a crer que usavam embarcações para navegação costeira. Estes grupos deixaram profundas marcas de seus assentamentos no litoral – os sambaquis – sítios arqueológicos formados predominantemente de conchas, mas compostos também por ossos de peixes e mamíferos (Rath, 1871; Capanema, 1874; Marques, 1880; Lacerda, 1885; Calixto, 1895; Ihering, 1898, 1903; Abreu, 1932; Gaspar, 2000).

Os sambaquis se apresentam como colinas que podem atingir, em alguns casos, altura de até 30 metros. O que diferencia os sambaquis dos concheiros naturais (assim denominados por muitos pesquisadores do início do século XX (Penna, 1876; Wiener, 1876; Hartt, 1885; Lofgren, 1893, 1903; Krone, 1902, 1908; Leonardos, 1938) é a presença de sepultamentos, vestígios de fogueiras, restos animais (e.g. dentes e ossos) e instrumentos (e.g. pontas de flechas e arpões) confeccionados pelos grupos sambaquieiros.

Foram construídos preferencialmente nos ambientes lagunares e estuarinos – considerados um dos locais mais férteis do mundo – ricos em moluscos, crustáceos e peixes. Vários povos indígenas, atraídos pela abundância de recursos marinhos, aí se estabeleceram e permaneceram por longo período, até que foram substituídos por sucessivas levas de povos agricultores e ceramistas, vindos do interior.

Os sambaquis são freqüentes nos litorais de todos os continentes, mas não se pode dizer que se tratava da mesma população, com a mesma unidade biológica e cultural. De fato, eram diversos grupos humanos que exploraram o mesmo ambiente, contando com a mesma matéria prima para a confecção de seu mobiliário, adaptado a necessidades semelhantes (Bartorelli e Isotta, 1965; Isotta, 1968). A literatura especializada em sambaquis converge para definição de diferentes sistemas que se construíram e reconstruíram como resposta a tensões e pressões multivariadas, ao longo de seis milênios (Piazza, 1974; Gaspar, 1991; Lima, 1999). Daí as configurações bastante distintas, em seus detalhes, dos vários sambaquis já estudados. A questão mais recente abordada pelas pesquisas trata do processo de formação destes sítios e a sua concepção como marco espacial (Gaspar e De Blassis, 1992; Afonso e De Blassis, 1994; Gaspar, 1995; Barbosa, 1999).

Entretanto os sítios insulares possuem composição diferente dos sambaquis (tradicionalmente conhecidos como um amontoado de conchas), pois a concha não foi o principal material construtivo para os "mounds" que comumente descritos. Porém, podemos considerar que sejam similares ao sistema sócio-cultural sambaquieiro.

A composição básica destes sítios é a terra preta, ossos de peixes e mamíferos marinhos e terrestres, sepultamentos humanos e muitos artefatos ósseos e líticos. As dimensões também são bem menores que os sambaquis, chegando ao máximo 2 metros de altura e diâmetros de 5 a 20 metros. As datações variam entre 5000 anos BP a 75 anos BP. Sua localização está, em sua maioria, na média vertente, nas ilhas e áreas costeiras.

Isto pode significar que a ilha pode ser, como uma paisagem natural, definida como um pedaço de terra rodeada de água, com um significado de uma área isolada, como refúgio, proteção, podendo estar tanto nas ilhas como no continente. Isso pode ser caracterizado também na área costeira, onde as pequenas enseadas rodeadas pelos morros ou com a Serra do Mar por trás da planície costeira aparecem como uma grande "ilha" pode ser observada a partir do mar ou de outro ponto de vista. Objetivamos neste trabalho entender a ocupação desses grupos pré-históricos nas ilhas e nessas "ilhas" das áreas continentais, sob a perspectiva da Arqueologia da Paisagem.

HISTÓRICO

Sambaquis da Baixada Santista

O Sambaqui de Maratuá, foi o primeiro a ser escavado na Baixada Santista sob a coordenação de Emperaire, com o patrocínio parcial do Musée de L' Homme de Paris, na pessoa de Paul Rivet que, entusiasmadíssimo com a riqueza dos sambaquis do litoral paulista auxiliou na fundação de um centro de pesquisas arqueológicas em São Paulo. Inicialmente com o nome de Comissão de Pré-História e posteriormente, em 1962, como Instituto de Pré-História da Universidade de São Paulo, hoje, compondo o Museu de Arqueologia e Etnologia.

O Sambaqui de Maratuá emergia de um mangue, à margem esquerda do rio Maratuá e já se encontrava destruído em cerca de dois terços quando ali iniciaram as pesquisas; mesmo assim, forneceu riquíssimo material ósseo humano, além de artefatos em ossos de animais e líticos (Fig. 13.1). Até hoje, foi o único sambaqui paulista onde foram encontradas duas finas agulhas com fundo, feitas em osso de peixe. Foi datado, inicialmente, em 7.803 ± 1.300, pela medição do C^{14}, na França e esteve envolvido em uma polêmica a cerca da cronologia dos sambaquis e sua relação com a oscilação do nível do mar para a Baixada Santista, foi redatado por Caio Del Rio Garcia, que concluiu por cronologia da ordem de 4000 anos antes do presente.

O Sambaqui de Mar Casado, datado de 4.400 ± 130 (GIF-1194) foi o segundo a ser escavado na Ilha de Santo Amaro, localizado próximo à Praia de Pernambuco, à altura do Km 3,5 da estrada Guarujá – Bertioga, na cabeceira do Rio Perequê, a uma altitude de 4,5 metros acima do nível do mar, hoje cerca de 400 metros de distância da praia (Fig. 13.2). A uma distância de 100 metros deste, havia ainda outro pequeno sambaqui, inteiramente destruído pela retirada de conchas, moídas para servirem de adubo para uma plantação.

A pesquisa iniciada em 1961, realizada sistematicamente e com o possível rigor científico durou um ano e dois meses, coordenada por Paulo Duarte e financiada pelo Conselho Nacional de Pesquisa e pela CAPES, do Ministério da Educação. O material coletado consistiu em belíssima e diversificada indústria óssea, de dentes de animais e lítica, além da exumação de oito indivíduos, representados por esqueletos incompletos e nenhum crânio. Entretanto, em 1948, Biocca, Schreber (biologistas) e Hoge (zoólogo) que escavaram parcialmente este sítio (já haviam feito algumas intervenções em Maratuá, também), encontraram restos de três indivíduos, inclusive uma criança e um crânio de adulto, os quais foram depositados, na época, no Museu Paulista, e, com a transferência do acervo em 1989 para a Cidade Universitária, hoje, fazem parte das séries esqueletais do MAE/USP.

São encontrados, ainda, referências a 26 crânios coletados nos "sambaquis da Ilha de Santo Amaro", ofertados por Richard Krone ao Museu Paulista e descritos por Mendes Corrêa em 1946 e por Willems e Shaden em 1951, também estes integrados ao acervo do MAE/USP. Em 1964, Niéde Guidon descreve as técnicas empregadas em campo e inventaria detalhadamente o material coletado, publicando junto com Luciana Pallestrini um estudo sobre a indústria óssea e de dentes.

O terceiro sambaqui escavado sistematicamente na Ilha de Santo Amaro foi o de Buracão nos anos de 1962 e 1963. Localizado junto ao canal de Bertioga, teve a pesquisa coordenada por Paulo Duarte, com a participação das arqueólogas citadas no parágrafo anterior, que iniciaram a investigação com o encaminhamento de novos problemas, levantando novas hipóteses, questionando e reinterpretando a função dos artefatos encontrados. O sitio havia sido quase completamente destruído pela abertura da estrada Guarujá – Bertioga, tendo sido explorado também por Biocca, Sheireiber e Hoge em 1947 que, na época o designaram de S3. Foram coletadas, além das séries esqueletais, variada indústria lítica e óssea. Em 1964, Luciana Pallestrini publicou as escavações e um inventário parcial dos artefatos coletados, descrevendo os

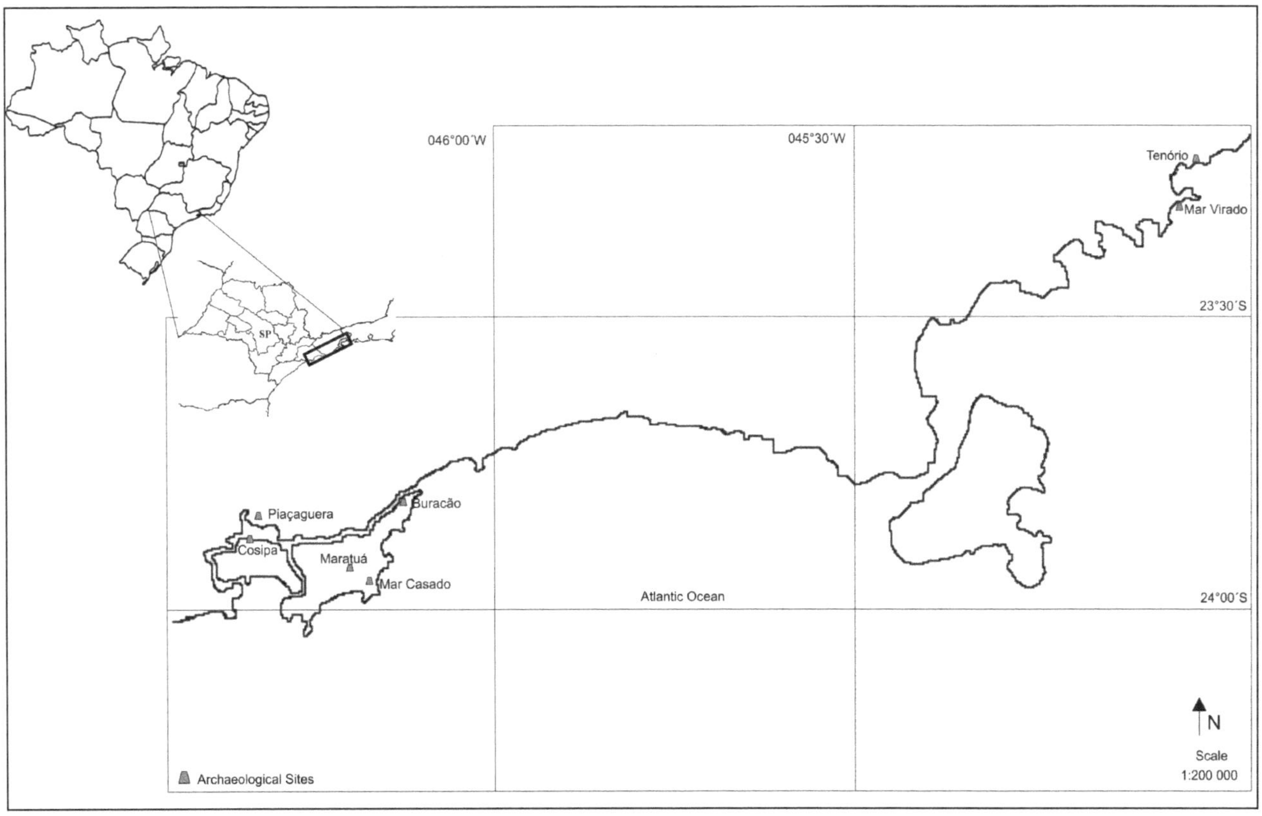

Fig. 13.1. Mapa da região norte-sudeste do litoral de São Paulo. Vista geral da localização dos sítios trabalhados

Fig. 13.2. Escavações realizadas no sambaqui do Mar Casado.
Foto: Documentação do Museu de Arqueologia e Etnologia

sepultamentos e seu mobiliário funerário. Buracão é diferente de Maratuá e Mar Casado que levam a crer se tratar de "sítio acampamento" ou "acampamento conchífero" e não de sambaqui como classificado anteriormente. Datado de 1950 ± 100 BP, atualmente apresenta sua série esqueletal incluída nas coleções que estão sendo analisadas por Sérgio Francisco Serafim Monteiro da Silva, que pretende identificar suas semelhanças/ diferenças com o material de sambaqui, por meio de variáveis culturais e biológicas.

Esses sambaquis de São Paulo, atualmente, localizam-se a 400 metros de distância da praia (Duarte, 1968) e, apesar do nome (Ilha de Santo Amaro) e de sua interligação com a área costeira, não apresentam o isolamento como uma "ilha" em si, porém temos as informações de Ab'Saber (apud Duarte, 1968) a respeito dos estudos geomorfológicos dessa área onde ele conclui que "a ilha de Santo Amaro fazia parte outrora de um arquipélago, o chamado Páleo-Arquipélago Santista, sendo diversas ilhas ligadas depois pelo trabalho incessante, contínuo de sedimentação eólica e marinha. Maratuá, achava-se talvez numa pequena ilha, Buracão em outra e Mar Casado em outra..."

Esta constatação é um dado muito interessante na medida em que o sítio Buracão (com a datação próxima ao do sítio Tenório) apresenta características morfológicas e culturais semelhantes às do sítio Tenório e do Mar Virado. Segundo Monteiro da Silva (informação oral, 2004), as características bioantropológicas do sítio Buracão e sua cultura material se aproximam dos dois sítios do município de Ubatuba (os resultados serão apresentados em sua tese de doutoramento). Além disso, temos a presença do pendente em concha de *Strombus* sp que só aparece nesses dois sítios também.

Mais uma vez, a questão da percepção da paisagem a respeito de um local isolado (quando observamos próximo ao local) e ao mesmo tempo um destaque na paisagem (os morros onde localizam esses sambaquis, quando estamos observando de um ponto mais afastado) se repete no caso do sítio Buracão.

Em 1963, a equipe do Museu Paulista, sob a coordenação de Luciana Pallestrini, iniciou as escavações no Sambaqui de Piaçaguera, localizado em terrenos da Companhia Siderúrgica Paulista (COSIPA). Esta iniciativa (não concluída) foi retomada pela equipe do Instituto de Pré-História da Universidade de São Paulo, sob a direção de Paulo Duarte e, posteriormente, Caio Del Rio Garcia, até sua conclusão em 1969. O sambaqui havia sido descoberto durante os trabalhos de terraplenagem, por ocasião das obras de instalação da COSIPA, quando um trator dividiu o sítio praticamente ao meio. A pesquisa foi de salvamento, pois o sítio estava em plena área industrial, em vias de destruição e, por isso, precisou ser totalmente escavado. Datado em 4930 ± 110 anos BP e rico em vestígios esqueletais, deu margem a vários estudos bioantropológicos e sobre a arqueologia da morte, dado a seu abundante e diversificado acompanhamento funerário, além de estudos de demografia. Foram analisados, também, seu material lítico e os restos faunísticos, cujos trabalhos foram os precursores da implantação dos estudos de zooarqueologia no país.

No entanto, é no decorrer dos anos 80 que a pesquisa em sambaquis na Baixada Santista se intensifica, com a elaboração de amplo projeto, interdisciplinar, interessado também na recuperação e preservação do patrimônio ambiental e cultural, abandonando assim, a era da pesquisa arqueológica isolada.

Sob a coordenação de Caio Del Rio Garcia e Dorath Pinto Uchoa, com total apoio da COSIPA é elaborado em 1982 o "Projeto COSIPA/USP: Preservação arqueológica, ecológica e histórica da Ilha do Casqueirinho, Cubatão, SP, Brasil", ocasião em que é celebrado convênio entre a COSIPA e a USP. Com base neste projeto, pretendia-se implantar na Ilha, o Parque do Casqueirinho, onde se desenvolveriam três programas principais: o de arqueologia pré-histórica, o de arqueologia histórica e o programa ecológico, de preservação, recuperação e manejo ambiental. A administração do Parque ficaria sob a responsabilidade da COSIPA. Neste novo programa constava ainda, a manutenção de parte de um sítio como monumento, a preservação de testemunhos arqueológicos em outros e designação de um dos sítios para "sítio escola". A recuperação das ruínas da caieira ficaria à cargo da COSIPA, sob a orientação do Conselho de Defesa do Patrimônio Histórico, Arqueológico, Artístico e Turístico – CONDEPHAAT. Para as escavações no entorno das ruínas foi convidada Margarida D. Andreatta do Museu Paulista da USP.

Os sambaquis e a caieira foram escavados em diversas etapas de campo, ocasião em que foram coletados artefatos, restos alimentares, os quais foram utilizados para estudos de problemas associados à zooarqueologia, dieta e distribuição de além da exumação de restos humanos em conexão anatômica e acompanhamento funerário.

A Ilha apresenta o testemunho de duas fases de ocupação humana; os sítios tipo sambaqui e a presença dos portugueses, representada pelo forno de cal, que utilizaram como matéria prima as conchas do Sambaqui COSIPA-1, implantando nos séculos XVII ou XVII a primeira "indústria" da ilha.

Os sítios datam de: 4210 ± 90 BP o Cosipa-1; 1180 ± 60 BP o Cosipa-2; 3790 ± 110 o Cosipa-3 e 2590 ± 80 BP o Cosipa-4, sugerindo que estes grupos permaneceram na Ilha por um período de mais de 1000 e menos de 5000 anos. No sub-projeto: Programa de recuperação e manejo ambiental, um grupo de biólogos com diversas especialidades traçam um plano de recuperação ambiental baseado no levantamento faunístico e florístico da Ilha do Casqueirinho, o qual, entretanto, não chegou a ser implantado.

Sítios do Litoral Norte

A paisagem do litoral Norte difere com o litoral centro-sul de São Paulo por apresentar um relevo recortado onde a Serra do Mar mergulha diretamente no mar, dando um aspecto de relevo afogado. Os cordões litorâneos desaparecem e formam pequenas enseadas com praias de pequenas extensões. As escarpas da Serra do Mar formam a linha da costa com inúmeras ilhas, e se "apresenta com

Fig. 13.3. Vista geral da enseada da Ilha do Mar Virado, a partir do mar. Foto: Sandra Amenomori

uma sucessão de pequenas enseadas e planícies costeiras, a maioria das quais de pequena expressão".

Certamente, a abundância de moluscos nesta área não é a mesma das do litoral Sul de São Paulo ou de Santa Catarina. Os manguezais se limitam a pequenas áreas e não temos áreas lagunares. Mesmo assim, não encontramos nenhum vestígio de sambaquis nas planícies mais extensas de Ubatuba, Picinguaba ou Ubatumirim.

Em função deste tipo de paisagem, os sítios arqueológicos encontrados nas ilhas e as pequenas enseadas, desde o litoral norte de São Paulo até o litoral sul do Rio de Janeiro, apresentam composições estratigráficas diferenciadas dos sambaquis (basicamente composto de terra preta e ossos de peixe) quanto a sua inserção na paisagem e sua morfologia.

Poucos sítios foram sistematicamente estudados na medida em que muitos já se apresentavam totalmente destruídos e outros foram registrados, como os identificados por Cali (2000) em diversas ilhas da região de Ilha Bela, com aproximadamente 10 sítios cadastrados.

Nessas ilhas não há drenagens perenes, quanto à topografia, um dos sítios estava numa área plana junto a costeira da praia, dois partiam da costeira e subiam pelo morro em meia encosta. Quatro deles estavam aproximadamente entre 70 e 150 metros de distância da costeira da praia, em meia encosta dos morros e nos topos dos morros. Interessante lembrar que em três desses sítios estavam associados abrigos sob rocha, sendo que, em pelo menos dois deles haviam sepultamentos.

O sítio Mar Virado localiza-se numa pequena enseada desta Ilha, situado na face protegida de NE, próximo à desembocadura de um pequeno curso fluvial (Fig. 13.3). (COORDENADAS UTM: 23K 0484708/ 7393740)

A escavação do sítio arqueológico na Ilha do Mar Virado, sob a coordenação da Profa. Dra. Dorath Pinto Uchoa, do Museu de Arqueologia e Etnologia em convênio com a FUNDART, da prefeitura de Ubatuba, foi iniciada em dezembro de 1990.

Em princípio havia sido em caráter de salvamento, na medida em que houve inúmeras intervenções antrópicas e/ou naturais na área. O trabalho se restringiria apenas a escavar o que restou do sítio. No entanto, ao delimitar a área a ser escavada foi observado que o sítio foi superficialmente destruído, havendo uma profundidade média de 1,0m a ser escavada, com uma área total de 304 m^2.

Os 1111 artefatos do sítio Mar Virado apresentaram a predominância de pontas simples, com 315 unidades. Destas 315, 306 são de espinhos de nadadeira. Das 52 pontas biterminadas, 41 também são de espinhos de nadadeira. Foram contabilizadas 199 aparas de espinho de nadadeira.

A composição do sítio é formada basicamente de ossos de peixe e mamíferos marinhos, espículas de ouriço e poucas conchas com uma matriz sedimentar de terra preta.

De acordo com Monteiro da Silva (2001) até o momento foram evidenciados e exumados 43 sepultamentos entre enterramentos primários e secundários, e por vezes apresentavam-se calcinados. Observou-se ainda que carapaças de tartaruga e ossos de baleia estavam sobre alguns dos indivíduos, além de blocos de granitos dispostos ao redor desses enterramentos.

Fig. 13.4. Vista do sítio CV1 a partir do sitio CV2. Foto: Sandra Amenomori

As duas datações para este sítio: 2570 ± 70BP (Cal.2360 a 2060 BP) e 2640 ± 70BP (Cal.2480 a 2140 BP).

O sítio Tenório localizava-se a 100 metros da linha da costa, na praia do Tenório, município de Ubatuba, São Paulo, com as coordenadas UTM aproximadas: 23K 0493206/7406637.

O terreno era relativamente plano, arenoso e coberto por vegetação de gramíneas. Devido à proximidade da Serra do Mar, as planícies costeiras se encontram compartimentadas entre esporões da Serra, que penetram no mar, constituindo promontórios.

O Sítio Tenório foi escavado (área de 250 m^2) no período de 1969 a 1970 em caráter de salvamento, decorrente da implantação de loteamentos. Datado de 1875 ± 90 AP ou 75 d.C. (Garcia, 1972); Uchôa,1980); este sítio tem como principais características a terra preta com os lentes de vestígios faunísticos em toda a sua extensão.

Dos 3050 artefatos analisados, a predominância é das pontas simples, sendo que das 372 pontas, 301 são de espinhos de nadadeira. Das 279 pontas biterminadas, apenas 42 foram confeccionadas em espinho de nadadeira. Neste caso, predomino as pontas em osso de mamífero, com 168 unidades. As aparas apareceram em um total de 245. Se considerarmos que a apara é um rejeito das pontas biterminadas, existe uma desproporcionalidade em relação a essas pontas.

Foi encontrado um total de 88 sepultamentos, sendo estudados 28 (em conexão anatômica). A característica marcante dos sepultamentos é a grande quantidade de seixo ao redor do indivíduo. A maioria apresentava-se em posição estendida e semi-fletida e não foi observada sobreposição como no sítio Tenório (Monteiro da Silva, Comunicação pessoal, 2005).

O sítio Couves 1 localiza-se na meia encosta próxima à segunda enseada da Ilha das Couves (Fig. 13.4). A vertente apresenta 27% de declividade, à 14 m dos matacões graníticos que se assentam na praia (nível médio da maré). (Coordenadas UTM: 23K 0514647/7409660).

A ilha apresenta uma área de 64,5 hectares com duas pequenas enseadas com praia. Não existem cursos d'água doce, apenas córregos intermitentes e pequenas nascentes.

O Sítio Couves 1 foi encontrado por Amenomori em 2002, durante a etapa de prospecção arqueológica realizada na região de Picinguaba e as ilhas adjacentes, no município de Ubatuba.

O sítio encontra-se pelo menos 50% destruído, pois a abertura da trilha parece ter retirado a porção principal do sítio. O sítio foi datado em 2200 ± 40 anos BP (Cal.2330 a 2120 BP) – Beta 200404. A área total do sítio foi de 150 m^2, com profundidades que variavam de 0,50 a 1,40 metros. O material encontrado foi a fauna, líticos (principalmente quartzo), poucas conchas e ossos humanos queimados e esparsos. A análise foi qualitativa.

DISCUSSÃO

Percepção e Visibilidade

A percepção da paisagem sob a perspectiva dos grupos pré-históricos requer um extremo cuidado, pois partimos de vestígios materiais fragmentários e a carga simbólica passa pelo conhecimento das necessidades e desejos

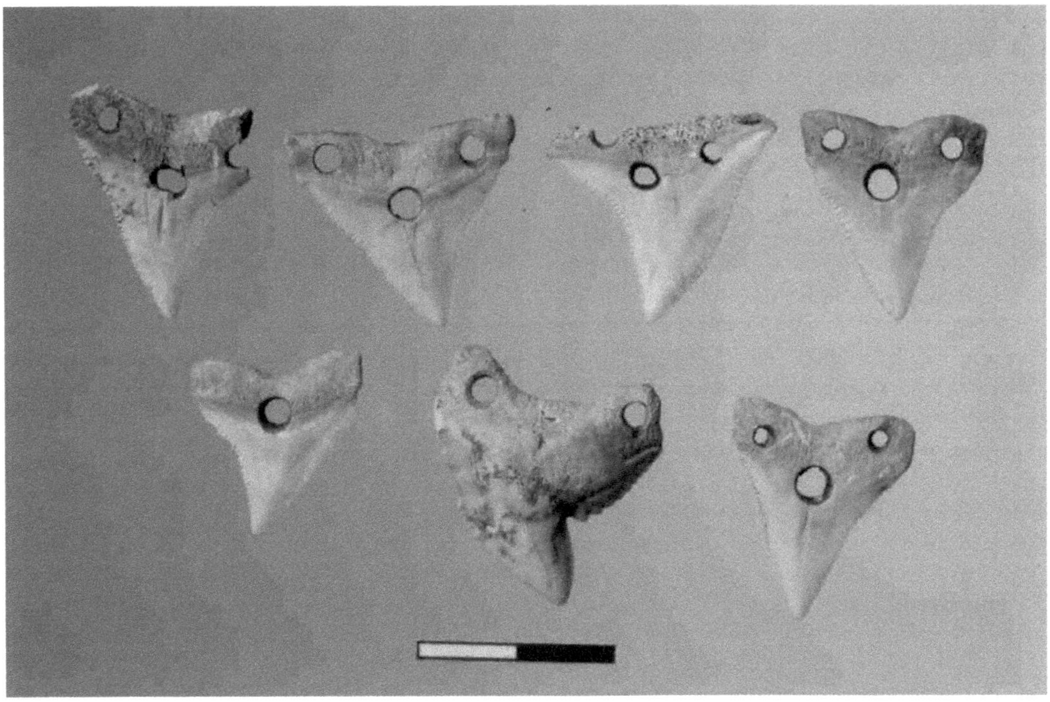

Fig. 13.5. Dentes de tubarão perfurados, acompanhamento funerário, Sítio do Mar Virado. Foto: Wagner Silva

humanos que estão manifestos na paisagem. Mesmo assim, partimos do conceito de que a percepção está associada aos vários sentidos que o ser humano possui, o olfato, tato, audição e visão; e nunca são utilizados isoladamente.

No caso dos grupos pescadores-coletores pré-históricos, a percepção da paisagem pode ser associada à própria localização dos sítios arqueológicos, onde observamos um padrão de assentamento, isto é, apresenta-se em áreas com pequenas enseadas, protegidas, muitas vezes localizadas na média vertente, sempre voltada para o continente.

No caso da área costeira do litoral norte do Estado de São Paulo, podemos acrescentar um dado em relação à circulação de ilha para continente e vice-versa, dos grupos pescadores-coletores pré-históricos, em que a visualização seria muito mais via marítima, isto é, observando a paisagem a partir de embarcações e não a partir da terra.

Inferimos a partir daí, que as embarcações eram muito mais utilizadas para circular entre as pequenas enseadas que são recortadas pela Serra do Mar do que andando pelas vertentes íngremes com a mata fechada. Esta constatação foi enfatizada pelas informações orais obtidas com os pescadores da Vila de Picinguaba, Praia de Camburi, e da ilha Montão de Trigo que disseram que há 40 anos atrás, o meio de transporte principal era os barcos. As estradas eram de terra e muito ruins, limitando o acesso das pessoas às áreas urbanas existentes na costa do litoral do Estado de São Paulo.

Este continuísmo de circulação que perdurou até a segunda metade do século XX prova que as embarcações tinham um papel fundamental no cotidiano das pessoas que viviam neste litoral recortado.

Não podemos esquecer que tais padrões de visibilidade e visualização dependem da escala de observação. A análise deve se basear numa aproximação do tipo "zoom", que permita reconstituir o contexto original por meio das relações espaciais significativas, como a análise formal ou morfológica (espaço físico ou construído ou a cultura material), a análise fisiográfica, a análise de trânsito (identificação das vias de comunicação), a análise das condições de visualização, visibilidade e intervisibilidade (ou relação visual entre esse elemento e outros, sejam arqueológicos ou não e, por último a análise de terrenos e análise topográfica).

Artefatos

Se os sítios arqueológicos estão localizados em uma determinada paisagem escolhida pelos antigos habitantes do litoral, a sua cultura material está embutida de significados para acompanhar esses grupos. Esses significados podem ser econômicos e ao mesmo tempo simbólicos. Ambos estão relacionados com a percepção do ambiente do seu entorno.

Os acompanhamentos funerários são muito mais elaborados nos indivíduos femininos do que nos masculinos, em ambos os sítios. O uso dos dentes de tubarões (Fig. 13.5), dos ossos de tartaruga, boto e baleia foi significativo em todos os sítios estudados. Esta recorrência do uso dos ossos desses animais marinhos como acompanhamento funerário, demonstra que os grupos pré-cerâmicos utilizavam-nas sob uma carga simbólica muito forte, dada a importância em utilizá-los como acompanhamentos funerários e que certamente a captura não foi realizada (com exceção das tartarugas) nessas áreas insulares. Provavelmente foram capturados e processados em outro local e trazidos com o intuito ritualístico.

Apesar de não podermos correlacionar diretamente os dados de sítios pré-históricos ao significado simbólico considerado em outras civilizações no mundo, as simbologias de certos animais denotam significados, muitas vezes, universais. Os golfinhos, na tradição chilena, aparece como peixes que transformavam em seres humanos pré-diluvianos, os quais, de tempos em tempos, saiam dos rios e vinham procriar com as mulheres.

Segundo o historiador Câmara Cascudo, alguns cronistas relataram histórias do boto como sendo a personificação do Uauiará, ou Uiara (Senhor das Águas), o grande amante das mulheres caboclas e índias na mitologia tupi. O primeiro filho de muitas nativas é atribuído ao contato com esse deus que, ora as surpreende no banho, ora transforma-se em mortal para seduzí-las, arrebatando-as para dentro das águas. O boto, como Uauiará, representa o variante masculino da Iara (Mãe-d'Água), dona de igual poder de encantamento e sedução. Assim, de modo amplo, o peixe está simbolizando o elemento água, dentro da qual vive. O peixe também é símbolo da vida e da fecundidade, em vista da sua prodigiosa faculdade de reprodução e do número infinito dos seus ovos. Em muitos mitos, por sinal, em várias partes do mundo, sempre houve referências à fecundação por deuses e entes mágicos.

As tartarugas, em vários países e regiões, como o Japão, China, África, Polinésia, significa longevidade e fertilidade. Na Índia, são símbolos da imortalidade e são consideradas a segunda encarnação do poderoso deus Vishnu na religião Hindu.

Na Polinésia, África negra e na Lapônia, a baleia em mitos iniciáticos análogos ao de Jonas – passam pelo ventre do monstro (marinho, com freqüência) é,às vezes, considerada expressamente como uma descida dos infernos.

Nas costas do Vietnã, os ossos das baleias encalhadas são recolhidos e constituem objeto de culto: divindade do mar, a baleia guia os barcos dos pescadores e salva-os dos naufrágios. Por simples extensão, o espírito-baleia apresenta-se também como um espírito bom e prestativo na passagem que leva à morada dos imortais.

Quanto aos líticos, apesar de não terem sido analisados sistematicamente, observamos que a matéria-prima utilizada para a confecção de várias lâminas de machado, lascas e outros artefatos, foi em seixos de rocha básica, isto é, basalto, traquito ou diabásio. Tais seixos foram coletados em outra localidade e levados para as áreas dos sítios, onde foram processados até chegar a forma do instrumento desejando. Pela grande quantidade de lascas e estilhas, além de núcleos no mesmo local dos enterramentos, podemos inferir que todos os instrumentos, seja ósseo como líticos foram confeccionados na mesma área dos sítios, onde foram constatada os vestígios de toda a cadeia operatória desses artefatos.

No sítio Mar Virado, Monteiro da Silva (2001) e Nishida (2001) já haviam discutido a respeito da bivalvia *Spondylus americanus*, cuja concha teria um significado simbólico pois só aparecia próxima aos enterramentos humanos, possivelmente como acompanhamento funerário.

A utilização de conchas, por exemplo, pode ter o significado de prosperidade, de sorte, como para os chineses. No Império Azteca, as conchas de *Spondylus* eram pintadas e transformadas em adornos e oferecidas aos deuses, cuja representação, significaria nascimento, geração, um útero.

Na Nova Guiné, Japão e outras ilhas, o som da concha *Strombus* servia para espantar os espíritos malignos, além de servir para o ritual de iniciação nos ritos sagrados.

Uma das conchas mais importante para os egípcios era a *Cypraea*. O nome científico *Cypraea* vem de Cyprus, a ilha onde nasceu Afrodite. Naquela época, as múmias tinham os olhos substituídos por Cypraeas, para que elas pudessem enxergar no "outro mundo". Também na China, conta Marco Polo que quando morria o imperador, sua boca era preenchida com nove Cypraeas, assim ele teria dinheiro para as despesas em sua nova vida. Em Pompéia, na Itália, a *Cypraea* conhecida como a concha de Vênus, era usada pelas mulheres como amuleto contra a esterilidade. Nas Ilhas Ryukyu, no Pacífico, próximo ao Japão, a mulher ao dar a luz, segura em cada mão uma *Cypraea tigris* para ter um bom parto. Esta concha simboliza, ainda hoje, o sexo entre as populações ribeirinhas do Oceano Índico e do Mar Vermelho (Sodré, 2005)

CONCLUSÃO

Os marcadores na cultura material podem ser apresentados como os artefatos junto aos sepultamentos, indicando um uso ritual e sua importância simbólica que podem ser inferidos quando se encontra em contexto funerário. O estar em contexto funerário significa uma aceitação ou relocação do mesmo por vias simbólicas: um machado ou adorno de pescoço ou de cabeça deixa de ser itens da vida cotidiana, de festas e caçadas e passam a fazer parte do simbolismo funerário. Quando não, foram exclusivamente elaborados pelos vivos para os rituais, para acompanhar o morto na cova e a sua nova "jornada". Assim, os acompanhamentos funerários podem ser, independentemente da matéria prima empregada para sua confecção, de origens e com propósitos diversos.

Esta percepção foi fundamental para os grupos que estavam confeccionando seus artefatos enquanto observavam o mar e pode ser entendida sob três aspectos – o primeiro é o controle de tudo e todos que se aproximariam da ilha, o segundo, é, o conhecimento da variação da maré para confeccionar as lâminas de machado, isto é, o grupo que habitou a ilha escolhia aleatoriamente os suportes rochosos utilizados, de acordo com essa variação da maré e o terceiro, mais subjetivo, haveria um sentido mais simbólico para a escolha de uma praia isolada e escondida.

Todas as representações neste trabalho estiveram de alguma forma, relacionadas com o mar. A pesca, identificada pelos trabalhos de zooarqueologia, indicou ser 90% retiradas do mar. Os caminhos e direções sempre foram a partir do mar. Os acompanhamentos funerários mais expressivos foram retirados do mar – o boto, a baleia, o tubarão e a tartaruga.

Os grupos de pescadores-coletores pré-históricos, que ocuparam as ilhas e áreas costeiras do litoral norte do estado de São Paulo, podem ser considerados como pequenos núcleos familiares (ou não) que tinham as áreas de circulação em torno de uma paisagem insular, onde realizavam suas práticas funerárias, sempre com uma percepção simbólica bem definida. A diversidade dos tipos de enterramento sugere pequenas variações entre esses núcleos, porém tinham a mesma percepção da paisagem – a ilha como um local isolado, um refúgio, seguro de tudo que se aproximava dela. Daí, a importância da visibilidade e visualização, pois os sentidos da visão e da audição eram primordiais para a escolha da paisagem insular.

Além disso, temos os artefatos que, apesar das semelhanças morfológicas entre a maioria dos instrumentos encontrados, alguns aparecem com características peculiares, podendo ser o identificador cultural de cada grupo (Tenório, 2003).

As práticas funerárias eram vinculadas às atividades cotidianas, como a pesca, a caça, a confecção dos artefatos e etc., isto é, todas essas atividades estariam relacionadas diretamente com o morto. Assim, podemos dizer que os grupos pescadores-coletores insulares apresentaram uma forte carga de simbolismo, tanto na maneira como se apropriavam da paisagem como no cotidiano para sua subsistência e práticas funerárias.

Bibliografia

ABREU, S.F. (1932) A importância dos sambaquis no estudo da pré-história do Brasil. *Revista da Sociedade de Geografia do Rio de Janeiro* 25.

AFONSO, M.C.; De BLASSIS, P. (1994) Aspectos da formação de um grande sambaqui: alguns indicadores em espinheiros ii , Joinville. *Revista do Museu de Arqueologia e Etnologia* 4: 21-30.

BARBOSA, M. (1999) Reconstituição especial de um assentamento de pescadores-coletores-caçadores pré-históricos no Rio de Janeiro. Pp. 205-221. *In:* Tenório, M.C. (Ed). *Pré-História da Terra Brasilis*. Rio de Janeiro, Editora da Universidade do Rio de Janeiro.

BARTORELLI, A; ISOTTA, A.L. (1965) Estudo do material lítico do sambaqui do Buracão. *Suplemento de Ciência e Cultura* 17(2):139. São Paulo, SBPC.

CALI, P. (2000) *Projeto Arqueológico em Ilha Bela, SP: Relatório de Atividades*. ano I.

CUNHA, S.B.; GUERRA, A.J.T. (1998) *Geomorfologia do Brasil*. Bertrand Brasil. p. 321.

CALIXTO, B. (1895-1902) Algumas notas e informações sobre a situação dos sambaquis de Itanhaen e de Santos. *Revista do Museu Paulista* 6:490-518.

CAPANEMA, G.S. (1876) Os sambaquis. *Ensaios de Sciencia* 1:79-91.

GAPAR, M.D. (1994) *Aspectos da organização de um grupo de pescadores, coletores e caçadores: região compreendida entre a Ilha Grande e o delta do Paraíba do Sul, Estado do Rio de Janeiro*. São Paulo: Tese de Doutorado, FFLCH, Universidade de São Paulo, 362p.

GAPAR, M.D.; De BLASIS, P.D. (1992) Construção de sambaqui. *Anais da Reunião Científica da Sociedade de Arqueologia Brasileira* 2:811-820.

GARCIA, C.D.R. (1972) *Estudo Comparativo das Fontes de Alimentação de Duas Populações Pré-Históricas do Litoral Paulista*. Tese (Doutorado). Instituto de Biociências, São Paulo, Universidade de São Paulo.128p.

HARTT, C. (1885) Contribuições para a etnologia do valle do amazonas. *Archivos do Museu Nacional* 6:1-174.

IHERING, H.V. (1898) Ueber die vermeinticle Errichtung der Sambaquis durch den Menschen. *Verhandlugen der Berliner Anthropologischen Gesselschaft* 217-222.

IHERING, H.V. (1903) A origem dos sambaquis. *Revista do Instituto Histórico de São Paulo* 8:446-458.

ISOTTA, C.A.L. (1968) O material lítico de sambaquis do litoral paulista. *Pré-História Brasileira*, 143-156.

KRONE, R. (1902) Contribuições para a ethnologia paulista. *Revista do Instituto Histórico e Geográfico de São Paulo* 7:470-482.

KRONE, R. (1908) Informações etnográficas do Vale do Rio Ribeira de Iguape. *In: Exploração do Rio Ribeira de Iguape*, São Paulo, Comissão Geográfica e Geológica do Estado de São Paulo, pp.23-31.

LACERDA, J.R. (1885) O homem dos sambaquis. Contribuição à arqueologia brasileira. *Arquivos do Museu Nacional do Rio de Janeiro* 6:175-257.

LEONARDOS, O.H. (1938) *Concheiros naturais e sambaquis*. Papéis Avulsos, Departamento Nacional da Produção Mineral – Serviço de Fomento da Produção Mineral 37.

LIMA, T.A. (1991) *Dos mariscos aos peixes: um estudo zooarqueológico de mudança de subsistência na pré-história do Rio de Janeiro.* Tese de Doutorado, Universidade de São Paulo, São Paulo.

LIMA, T.A. (1999-2000) Em busca dos frutos do mar: os pescadores- coletores do litoral centro-sul do Brasil. *Revista USP* 44:.207-327.

LOEFGREN, A. (1893) *Os sambaquis de S. Paulo.* Comissão Geographica e Geologica do Estado de São Paulo, 9.

MARQUES, A.A.S. (1880) As ostreiras de Santos e os kjokkenmoddings da Dinamarca. *Revista Brazileira* 6:185.

MARTIN, L.; SUGUIO, K.; FLEXOR, J.M. (1986) Shell middens as a source for additional information in Holocene shoreline and sea-level reconstruction: exemples from the coast of Brazil. *In:* Van de Plasseche, O (ed.) *Sea-Level Research: A Manual for the Collection and Evaluation of Data.* Amsterdam, Free University of Amsterdam, pp.503-521.

PIAZZA, W. (1974) Dados à arqueologia do litoral norte e do planalto de Canoinhas. *Publicações Avulsas* 26:71-86.

RATH, C.J.F. (1874) Die sambaquis oder Muschelhügelgräber Brasiliens. *Globus* 26:193-198.

ROSINI, A. (2001) Análise temporal do complexo estuarino da Baixada Santista – SP, por meio de imagens TM-Landsat. *Relatório Técnico INPE* p.1-13.

UCHÔA, D.P. (1969) Nota prévia sobre os sepultamentos do sambaqui de Piaçaguera. *Estudos de Pré-História Geral e Brasileira*, pp.487-492.

UCHÔA, D.P. (1980) Sinopse do "Arcaico do Litoral de São Paulo". *Anuário de Divulgação Científica 7* . Instituto Goiano de Pré-História e Antropologia. Universidade Católica De Goiás. Goiânia.

UCHÔA, D.P. (1982) Ocupação do litoral sul-sudeste brasileiro por grupos coletor-pescadores holocênicos. *Arquivo do Museu de História Natural* 7:133-143.

WIENER, C. (1876) Die sambaquis oder Muschelhausen in Brasilien. *Das Ausland* 45:895-897.

THE SUBMERGED SHELL MOUNDS OF *CANANÉIA, SÃO PAULO*, BRAZIL: A CASE STUDY OF UNDERWATER ARCHAEOLOGY

Flávio Rizzi CALIPPO

Collaborating Researcher of the Center of Studies of Nautical and Underwater Archaeology of the State University of Campinas (CEANS/NEPAM/UNICAMP) and Doctorade's student in the Museum of Archaeology and Etnology of the São Paulo University (MAE/USP)

***Abstract**: In the Coastal Plain of Cananéia-Iguape (extreme south of São Paulo's coastline), it was founded several evidences of shell mound builders that was dated back to almost 8.000 years ago and that can be associate to a region that was submerged. Aiming at the understanding of these shell middens occurrence and the contextualization in relation to the holocenic variations of the relative sea level, they were studied by underwater archeological methods and techniques, on the geoarcheological interpretation of Waters (1992) and on the studies of Formation Process adopted by Stein (1992).*

Thus, the shell middens preserving potential and distribution in this area shows a strong correlation with the sea related level changes that occurred after the last holocenic regressive maximum. These evidences also indicated that these sites are structures constructed with intentionality (dipped of monumentality and meanings), which the oldest ones can be preserved under or above of the actual sea level. Contributing to the development of the brazilian underwater archeology and for the studies of the sea level changes.

***Keywords**: Underwater shellmounds; underwater Archaeology; formation process; relative sea level changes*

***Résumé**: Dans la plaine côtière de Cananéia-Iguape (littoral sud de São Paulo, Brésil), on a trouvé des preuves de l'occupation des amas coquilliers de au moins 8000 ans BP. Une partie de ces sites peut être aussi liée au plateau continental actuellement submergé. Afin de les replacer dans leurs contextes en ce qui concerne les changements de l'environnement à l'Holocène, ces sites et cette occupation ont été étudiés appuyés sur les théories de l'archéologie subaquatique, de la géoarchéologie (Waters, 1992) et des études des processus de formation adoptés par Stein (1992).*

Cette recherche a pu constater que le potentiel de conservation et la distribution des amas coquilliers de cette région montrent une remarquable corrélation avec les variations du niveau de la mer. Au même titre que certains évidences indiquent que ces structures ont été construites avec d' intention par leurs habitants (imprégnées de signification et sites monumentaux) et les sambaquis (amas coquillers) submergés, les plus anciens, peuvent être conservés dans les milieux émergés aujourd'hui et aussi dans les milieux submergés.

Ainsi, cet étude contribue aussi pour le développement de l'archéologie subaquatique brésilienne et pour l'étude des variations du niveau relatif de la mer.

***Mots-clés**: Amas coquillers submergés; Archéologie subaquatique; processus de formation; les variations du niveau relatif de la mer*

***Resumo**: Em meio à Planície Costeira de Cananéia-Iguape (litoral sul do estado de São Paulo, Brasil) foram encontradas evidências de uma presença sambaquieira que remonta a pelo menos 8.000 anos AP e que pode ter parte de seus sítios associados a uma plataforma continental hoje submersa. A fim de contextualizá-los ao longo das modificações ambientais holocênicas, esses sítios e a ocupação sambaquieira que se estabeleceu nessa região foram pesquisados a partir de abordagens focadas na Arqueologia subaquática, na Geoarqueológica (Waters, 1992) e no estudo dos processos de formação adotados por Stein (1992).*

Além de identificar que o potencial de preservação e a distribuição dos sambaquis dessa região apresentam uma forte correlação com as variações do nível relativo do mar, algumas das evidências sugerem também que esses sambaquis são estruturas intencionalmente construídas (imbuídas de significado e monumentalidade), cujos representantes mais antigos (sambaquis submersos) podem ter se preservado tanto em áreas hoje emersas como submersas. Contribuindo, assim, também para o desenvolvimento da Arqueologia subaquática brasileira e para o estudo das variações do nível relativo do mar.

***Palavras-chave**: Sambaquis submersos; Arqueologia subaquática; processos de formação; mudanças do nível relativo do mar*

INTRODUCTION

Calippo (2004) proposed an initial model for the shell mound builders occupation of the Cananéia region, focused, mainly, in the environmental modifications that had occurred in consequence of the last great holocenic sea level change. Such model indicates that the arrival of the groups may have happened in two diferent – but non excluding – ways (Fig. 14.1): in the first one, the groups coming from the plateaus would have arrived at the littoral through the valley formed by the Ribeira de Iguape river and dominated an area that, probably, today, is submerged; in the second one, the migration would have happened through the coast, along, areas that actually are also re-covered by the sea.

In result of this direct association with the transgressive and regressive sea level movements, the research of the socio-cultural systems of the shell mound builders is progressively founded more on the relationship between the human groups and the environmental contexts where they were inserted.

In order to be able to discuss such relations, we chose the same approach as Stein (1992). The autor interpreted the context of the shell mounds of northwest coast of the

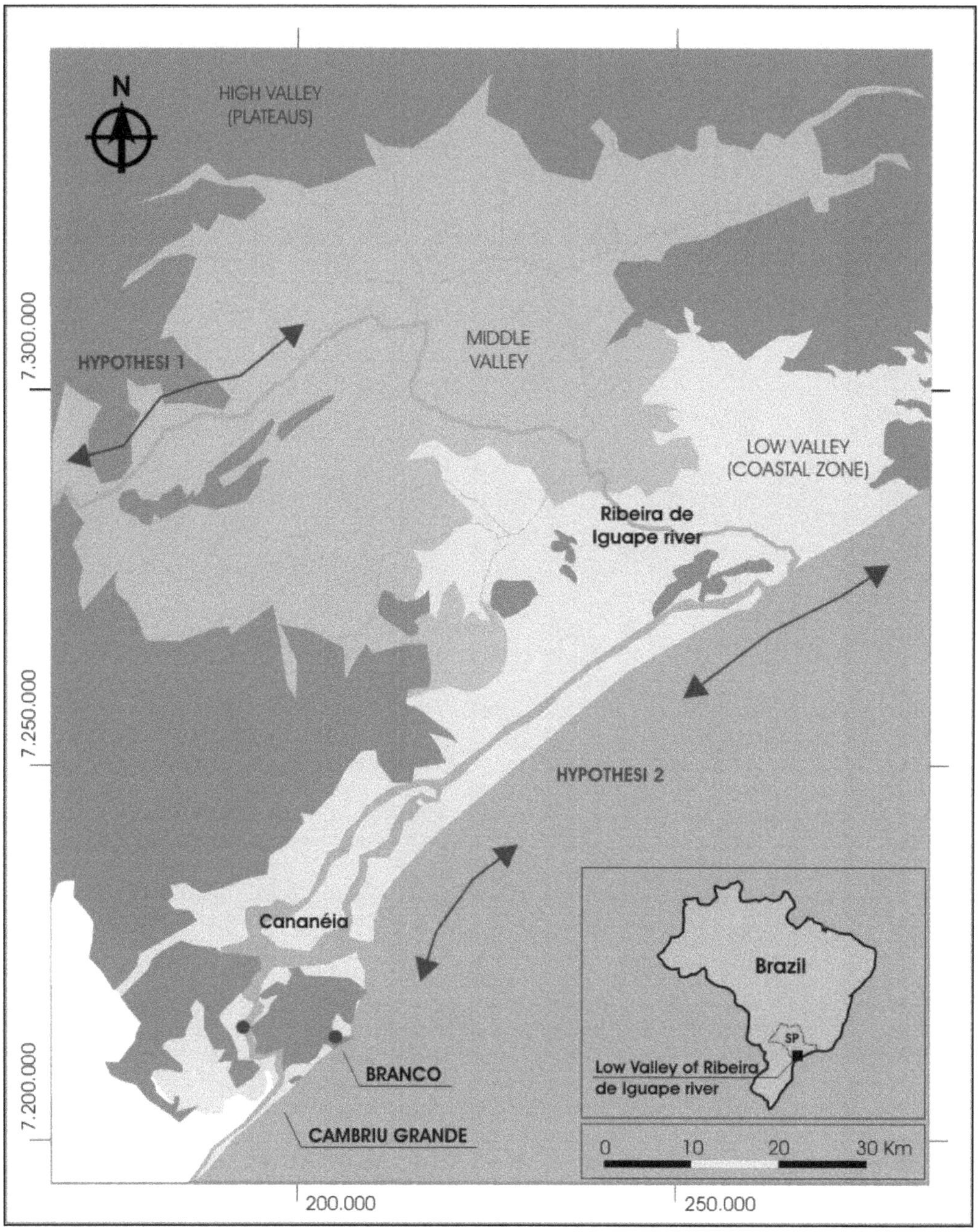

Fig. 14.1. Localization map of Low Valley of Ribeira de Iguape river and the shell mounds

United States, based on a geoarchaeological approach of formation processes presented by Schiffer (1972) and the Environmental Archaeology proposed by Butzer (1964).

THE RELATIONSHIP OF MOUNDS AND MOUND FORMATION TO THE SURROUNDING ENVIRONMENT (GEOARCHAEOLOGY ON THE BIGGER SCALE)

As important as it is to determine the origin of the Vale do Ribeira's humans groups, it is to understand that significant parts of the material culture of these people is today submerged. In this region we found the oldest shell mound of the Brazilian coast (Cambriu Grande), which is probably representative of an archaeological context still completely unknown. It is possible too, that there are other sites in the same conditions, today submerged along the coast or even covered by recent coastal sediments.

The Cambriu Grande (UTM 7216076 0204325) shell mound was dated at 7.870 ± 80 BP and probably still exists only because it was protected from the sea level rise by a great rocky outcrop and other topographic compartments (Fig. 14.3) which acted as a geological trap, avoiding the destruction of the site.

The age, localization, faunistic composition and the evolutionary model of the coastal plain indicate that this site was close to the lake and the beach, which probably

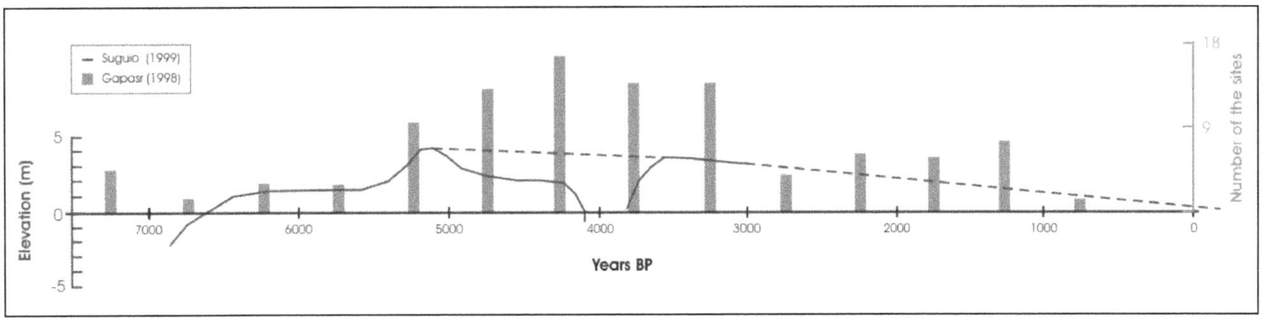

Fig. 14.2. Temporal distribution of Brazilian shell mounds and sea level changes

provide the resources for the construction of the site. These conditions seem to have lasted, aproximately, until the maximum holocenic transgression, when the sea entered the bay that borders it.

This trend can be verified when analyzing the values of the $^{12}C/^{13}C$ ratio present in the shells of this site: -0.7% and -0.1%, respectively, dated in 7.870 ± 80 and 5.390 ± 70 BP. Such values indicate an activity of collection associated to areas of intense, however, regressive, marine influences. Again, this suggests a strong relationship of the action of sea level change, since the most recent dating of the top of this site coincides with the holocenic transgressive maximum.

As important as the ancient age of this shell mound and the human presence in the Cananéia region, it is to perceive that the present spatial and temporal distribution of the Brazilian shell mounds may be just a part of a much larger archaeological context. In the Fig. 14.2 we can see that there is an inverse correlation between the number of the sites (Gaspar, 1998) and the sea level rise (Suguio, 1999).

When analizing the date of 7.000 years BP, an association between this small number of sites and the moment where the rising sea level moved past its present position for the last time cannot be avoided. Perhaps the small number of sites older than 5.500 years BP should be larger, for, if shell mounds older than these had been constructed, they would have been associated with shorelines currently found below present sea level. Probably, if sea level today was lower, further evidences of older sites could be found on emerged areas.

The minimal occurrence of shell mounds between 7.000 and 5.500 years BP can also be analyzed on the basis of sea level rise. It is possible that the small number of sites present in Fig. 14.2, is only part of all the shell mounds that were constructed, namely those on higher areas or those that, for some reason, had not been destroyed by the coastal agents. Such association seems to be confirmed when considering the sea level during the period of maximum occurrence of the shell mounds. When the sea level reached a standstill, the sea stopped destroying the sites and today we can see better preserved shell mounds.

THE REASONS FOR MOUND FORMATION AND THE INTERNAL STRUCTURE OF MOUNDS IN RELATION TO MICROSTRATIGRAPHY, INTERNAL STRUCTURES, ETC

In order to argue the reasons behind the formation processes of the shell mounds of the Cananéia region, we will present here the relative data to another site called Branco (UTM 7214409 0194260) (Fig. 14.4), which presents substantial evidences of alterations in its process of formation, probably associated with the change of function.

In order to investigate the formation process of this site twelve stratigraphic samplings were carried out using a vibracorer, being three of them in the submerged portion of the shell mound. The analysis and correlation of such samples indicated the existence of, at least, three different moments of formation, represented through a depositional sequence formed by three main sets of archaeological layers, namely Layer I, Layer II and Layer III.

The radiocarbon datings performed on samples from the top and the base of these layers indicate that these three changes occurred in an interval lasting not more 400 years. The first formed layer (Layer III) is probably relative to a short moment or perhaps to a single event of construction, because the difference between the dates of its top and base is restricted by the precision interval of the dating method. Differently of what it occurred in Layer II, which presented a difference of 300 years between its medium portion (4460 ± 60 years BP) and its top (4160 ± 50 years BP), the period of formation of Layer I was also restricted to the precision of the dating.

Despite the similarity between Layers I and III, there is a factor that cannot be ignored: Layer I is thicker than Layers II and III together. This, besides indicating that Layer I was also formed during an event of short duration suggests that such process could have occurred in a more intense way than the one associated with Layer III.

At a first moment, it is considered that the lowest layer is possibly associated with the formation of a base for the occupation of the site. This interpretation is corroborated by the rusty coloration and a higher degree of dissolution

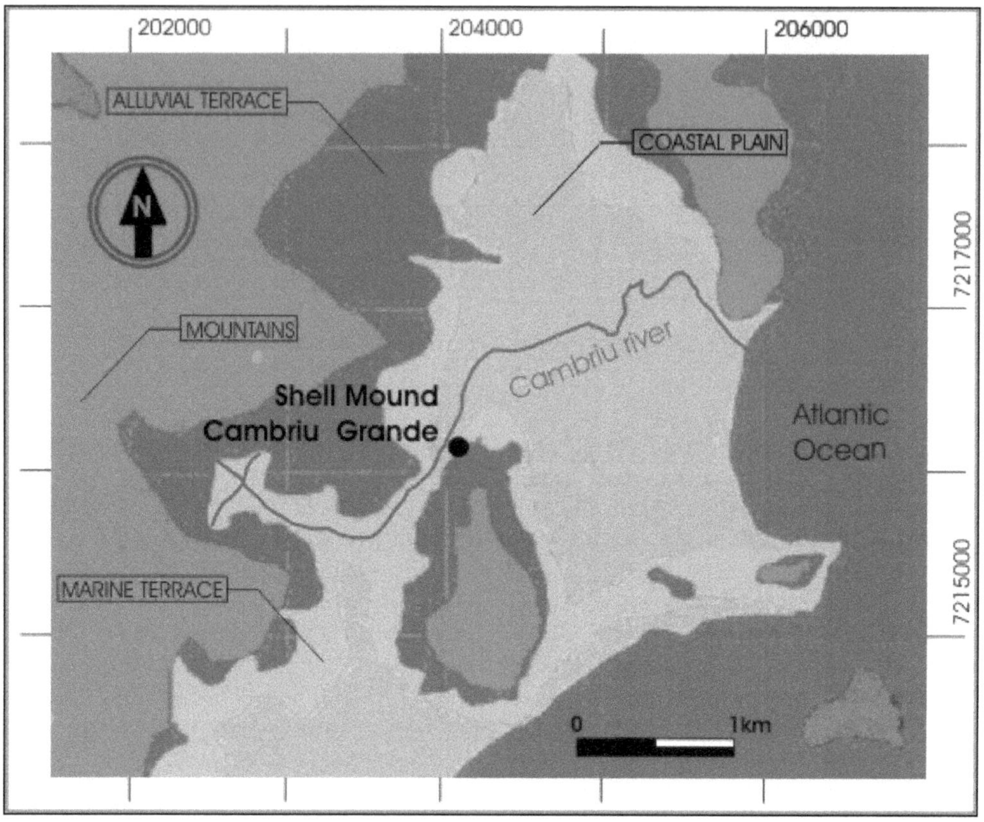

Fig. 14.3. The Cambriu Grande shell mound and its environment surrouding

of calcium carbonate of the shells. Since these processes are driven mainly by the chemical reactions of seawater and estuary water, there is the possibility that this layer acted as an isolation area, preventing the damp in the ground from reaching Layers I and II as well as aiding the drainage of the rain water. Those functions only became possible because of the largest grain size and porosity of Layer III, which restricts the retention and percolation of water by capillarity.

As Layer II is composed mainly of *valvae* of shells sufficiently fragmented, that uniformly distributed amid fine sediments, rich in organic matter, it is possible to infer that it must be resultant from a grater presence of humans at the moment of its construction.

According to De Blasis *et al.* (1998), this alternation of episodes of fast accumulation and depositional moments of long duration is found in the shell mounds of this region. The former generally composed of thicker, little compacted homogeneous layers (frequently shell from a single species), and the latter normally composed by fine layers of heterogeneous materials, including clays, sands and mixtures of fragmented shells. Most of the time these layers are mixed with fine lenses of coal (related to fires) and present a dark coloration.

Like Layer III, Layer I is probably the result of an episodical formation, because, as considered by De Blasis *et al.* (op. cit.), this type of formative process is verified by the presence of a single thick layer of shells that apparently doesn't have laminations or any other type of structure which would indicate chronic events of occupation. Besides Layer I being thicker than Layer III, the main difference between them, is the fact that it isn't rare to find closed bivalves in Layer I, indicating that these mollusks could have been collected and led to the shell mound mainly with the intention of raising the level of its surface.

To confirm this hypothesis we worked with the aid of the Laboratory of the Paleoceanography of the South Atlantic (LAPAS/IO/USP) to analyze the phosphate content and the oxygen and carbon isotopic composition of each of these layers. These analyses aims respectively at better understanding their nature (areas of discarding, of intense occupation, etc) and the original environments of the shells.

The evidences found, besides allowing the elaboration of an initial hypothesis regarding the process of formation of the Branco shell mound, also made possible a better understanding regarding the paper of the natural agents acting on the processes of its formation (including the postdepositional processes) and also of other shell mounds throughout the last great holocenic sea level change for the region of Cananéia.

Since at the Cananéia region the sea reached its Holocene maximum around 5.100 BP four meters above its present

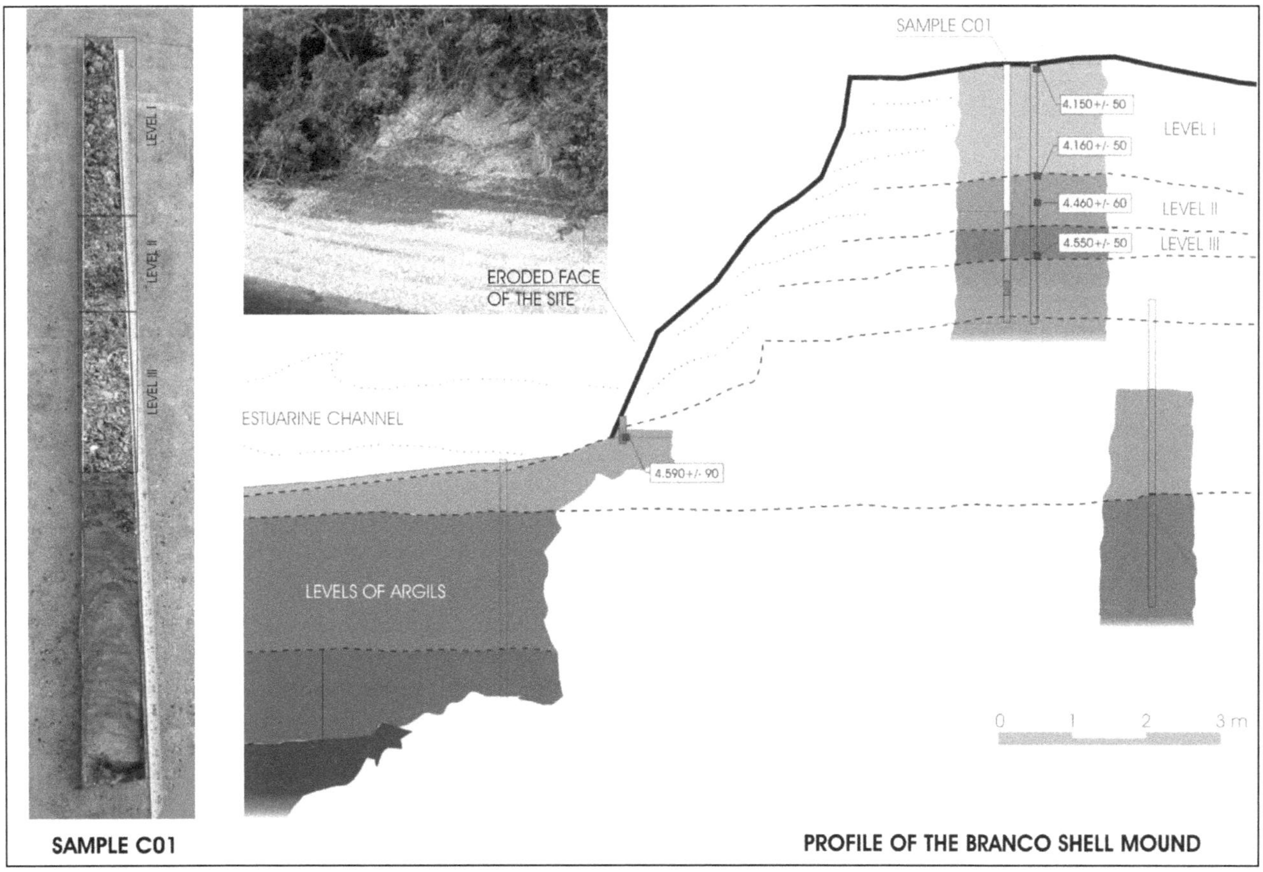

Fig. 14.4. Profile of the Branco shell mound

level and the oldest age of Branco shell mound is 4.590 ± 90 BP, we expected to find the base of this site at least four meters above sea level. However, this did not happened. The base of this site is very close to present sea level. There is one contradiction here for if this site is now near the sea level, it should have been about four meters deep when its construction began. But cannot be true either. Then, how could this shell mound have been constructed at a moment if the region where it is found was completely drowned? To attend this question, Angulo and Lessa (1997) suggest that as this region was formed by deposits of fine and hydrated sediments (Fig. 14.4), the level of this site decreased mainly because of subsidence processes, resulting from the compaction or loss of volume of these deposits in consequence of drying processes.

FINAL CONSIDERATIONS

When we begun this research we thought that all submerged shell mounds were those sites that simply had been submerged by the sea level and that the could be, at any time, found as they were. In spite of archaeological and geologic evidences showing the existence of destructive coastal agents, we were surprised to find that the shell mounds not always finished being destroyed or submerged. In some cases, archaeological evidences that were supposed to be submerged, are found in land, covered by recent coastal sediments or protected by some geological formation.

Analyzing these facts it was possible to understand some of the variables that constitute one of the main contributions of this research: the perception of the influence of the holocenic events of relative sea level change on the process of formation and the distribution of the shell mounds in space and time. As now this influence cannot be simply rejected, it is necessary to think about how to find them throughout the Brazilian coast.

The solution founded to this research, mainly, in what it refers to the first moments of the human occupation of the low valley of the Ribeira de Iguape river, was the elaboration of a Predictive Model. As well as already considered for the continental shelves of the United States (Waters, 1992), such models will allow us to identify the possible places of occurrence of submerged sites, the relative potential of preservation for each one and, mainly, the methods and techniques that will have to be used for their localization.

In this process, concepts and strategies will have to be reviewed. Perhaps, as it is already being done throughout the Australian coast (Bonhomme, 1999), some kinds of natural shell deposits will have to be investigated as probable destroyed sites. In collaboration with oceanographical and geological projects, we need to examinate the seafloor in search of older shell mounds, as it happened on the coast of South Carolina (U.S.A.).

References

ANGULO, R.J., LESSA, G.C. The brasilian sea level curves: a critical review with emphasis on the curves from Paranaguá and Cananéia regions. *Marine Geology*, 1997.

BONHOMME, Theresa. Shell middens on the central coast of New South Wales: A discussion of problems in distinguishing natural Shell deposits from cultural Shell accumulations. In: Hall, Jay; Mcniven, Ian J. (Org.). *Australian Coastal Archaeology*. Canberra, Australia: ANH Publications, Department of Archaeology and Natural History, RSPAS, the Australian National University, 1999, p. 229-233.

BUTZER, K.W. *Environment and Archaeology: an introduction to Pleistocene geography*. Aldine, Chicago: 1964.

CALIPPO, F.R. *Os sambaquis submersos de Cananéia, SP: um estudo de caso de Arqueologia Subaquática*. Dissertação de Mestrado. Faculdade de Filosofia, Letras e Ciências Humanas da Universidade de São Paulo. 2004.

DE BLASIS, P.A.; FIHS, S.R.; GASPAR, M.D.; FISH, P.R. Some references for the discussion of complexity among the sambaqui moundbuilders from the southern shores of Brazil. *Revista de Arqueologia Americana*. Instituto Panamericano de Geografia e Historia, v. 15, p. 75-106 jul./dez., 1998.

GASPAR, Maria Dulce. Considerations of the sambaquis of the brazilian coast. *Antiquity*, v. 72, n. 277, p. 592-615. 1998.

SCHIFFER, Michael B. Archaeological context and systemic context. *American Antiquity*, 37, p. 156-165, 1972.

STEIN, Julie K. *Deciphering a shell midden*. Seattle, Washington: University of Washington. Academic Press, Inc. 1992.

SUGUIO, K. *Geologia do quaternário e mudanças ambientais (passado + presente = futuro?)*. São Paulo: Paulo's Comunicações e Artes, 1999.

WATERS, M.R. *Principles of Geoarchaeology: a north american perspective*. Tucson: University of Arizona Press, 1992.

"MICROSTRATIGRAPHY OF SHELL MIDDENS OF TIERRA DEL FUEGO"

A. VILA, M. MADELLA, D. ZURRO, I. CLEMENTE, X. TERRADAS, I. BRIZ
Departament d'Arqueologia i Antropologia. IMF- CSIC Barcelona

J. ESTÉVEZ, J.A. BARCELÓ, E. VERDÚN, R. PIQUE, L. MAMELI
Grup de recerca d'Arqueologia Social americana. Universitat Autònoma de Barcelona. Unitat Associada al CSIC

E. PIANA
Antropología. CADIC – Ushuaia

Abstract: *The results of the joint research of Spanish and Argentinean teams since 1986 on shell middens of Tierra del Fuego can be useful guidelines for excavating and explaining similar archaeological sites around the world. We assume that human activity is organized in space and time and, as a consequence the random, non-random or homogeneous structure the resulting archaeological evidence has to be demonstrated before assuming a palimpsest nature of the deposit. We show the possibilities and relevance of recovering microstratigraphy of shell middens and discuss the need of using sampling methods that cover a wide surface of short discrete occupation episodes.*
Key words: *Shell middens, Microstatigraphy, Sampling methods*

Résumé: *Après plus de vint ans de recherches au Canal Beagle (la Terre du Feu) nous croyons utile de présenter nos résultats à propos de la méthodologie de fouille des gisements archéologiques de la zone, gisements appelées amas de coquillages ou "concheros". C'est l'essai de validation de l'hypothèse sur le caractère palimpseste de ces amas qui nous a conduit a cette méthodologie. L'application systématique de l'ensemble des techniques nous a permis voir l'importance de récupérer la microstratigraphie et développer un système de échantillonnage extensif sur des surfaces d'occupation.*
Mots clés: *Amas de coquillages, Microstratigraphie, systèmes d'echantillonnage*

THE DEVELOPMENT OF AN EXCAVATION AND RECORDING SYSTEM

"Shell middens", "amas de coquillages", "concheiros", "sambaquis", "escargotiers", "concheros", "conchales", "kjokkenmoddings", "kitchen middens"...There are many names in different languages just to refer to a sort of archaeological site characterized by high accumulations of shells and that denotes the attention paid to them. They are very variable in shape and size and were also attention calling for their high archaeological visibility and the very good preservation of the archaeological remains contained (cf. Waselkov, 1987).

However, the character of undifferentiated palimpsest commonly attributed to shell mounds has reduced these advantages of good preservation of the archaeological record. The usual excavation systems tend to segment the deposits in coarse arbitrary partitions of the stratigraphy and sample the record in columns assuming an overall homogeneous distribution of the remains.

Actually, shell middens include very often a complex intercurrence of different nature lenses (Orquera and Piana 1992) mainly because of:

1) The high-speed volume accumulation due to the large amount of residues produced by shellfish consumption and the amount of air contained before being compacted.
2) The limited extension of circumscribed deposits.
3) The scarcity of soil matrix that imbeddes the archaeological remains -shells included- which enables to distinguish short abandon moments that are not so easy to notice in other types of archaeological sites.
4) The possibility of being used as trash accumulations and/or for direct occupation; this also enlarging the possibility of anthropogenic postpositional alterations.

This complexity (Orquera and Piana 1992, 2000 and 2001) imposes the need to develop a specific methodology for the analysis and interpretation of archaeological evidences.

Archaeologically the sites on the coast of Tierra del Fuego have been described as shell middens. They are consequence of the social activities of human groups that have intensively exploited the littoral and maritime resources including a large amount of mollusks (basically mussels).

In this paper, our aim is to present some methodological issues as developed by the joint research of Spanish and Argentinean teams since 1986 on the Archaeology of the southernmost part of America.

It is our idea that such issues can be useful as general guidelines for excavating and explaining similar archaeological sites in other parts of the world. First, we assume that human activity is organized in space and in time. Therefore the resulting archaeological evidence must be, on principle, organized. In any case, the random, non-random or homogeneous structure of the evidence has to be demonstrated before assuming a palimpsest nature of the deposit.

Here we will focus only on the possibilities and relevance of a methodology trying the microstratigraphy of shell middens and consequently discussing the need of using a sampling method that covers a wide surface of short discrete occupation episodes.

First Essays: Lancha Packewaia and Túnel I sites

The Argentinean researchers were interested in the history and variability of fisher-hunter-gatherer human adaptation in the Beagle Channel, from the peopling of the area to the ethnohistorical *Yámana* people (Orquera *et al.* 2006) but before the joint research had not studied any archaeological site from the ethnohistorical period. The Spanish research team was interested in the development and testing of new methods and in the theory about hunter-gatherers archaeology, specially focusing in Social Archaeology subjects (Vila, 2006, 2004; Vila and Wünsch, 1990).

Consequently, since 1986 a Spanish-Argentinean team worked on eighteen and nineteen century archaeological sites of the northern coast of the Beagle Channel, this zone was, according to the descriptions of the ethnographers and travelers, part of "territory" of *Yamana* people (Estévez and Vila, 1995).

Archaeological systematic studies in Tierra del Fuego started with the *Proyecto Arqueológico Canal Beagle* (PACB) in 1975 (Orquera *et al.* 1978 and Orquera y Piana 1988, Orquera *et al.* 2006). According to the normal practice of that time, the first excavations (Lancha Packewaia and the beginning Tunel I sites) were done by artificial levels within broad natural layers, even though acknowledging –a simple look at shell middens profiles makes it obvious- that such technique will most probably mix different occupational events.

The search for a better excavation method to record and recover the stratigraphic complexity, the occupational sequence and the formation processes of the shellmiddens led the PACB, in 1982, to subdivide the shell layers into sub-shell middens. At the beginning the essay-error process was guided by observable criteria such as shell's state, consistence, texture, disposition and color, the soil matrix nature, abundance and color and the presence or absence of other elements as small pebbles (Orquera and Piana 1992). Soon, a more trustable independent variable was found: sub-shell middens were separated from each other by stratigraphic surfaces that were possible to detect at the excavation. This allowed isolating a sequence of subunits as if they were book pages. As it is explained in Orquera and Piana (1992), shell middens are a pack of interlocked shells separated from other units mostly by stratigraphic discordant surfaces, and occasionally by very thin non-shells layers. An experimental study on sub-shell middens formation was started in 1984. The results enabled to understand how such stratigraphic surfaces were formed, therefore this excavation methodology was tested in Tunel I and other archaeological sites as *Shamakush I* and *Shamakush X*.

Since the recovering of such differences and the recognition of criteria depends on the digger's recognition capability, there was a need for a strict and detailed recording, complemented by an exhaustive photographic control and a later laboratory re-analysis. So, other than tridimentional registering of the findings, also each subunit pre-excavation limits was measured and pictured before and after it's extraction.

Still an effort was made to develop a better recording method.

From 1988 in the frame of our Spanish-Argentinean international research projects,* the sampling techniques have been developed and new registering methods have been used.

The sites Túnel VII and Lanashuaia: standarization of the recording methods

This new way of excavating an archaeological shell midden was systematized in a registration protocol at the Spanish-Argentinean excavations of *Túnel VII*, a yamana settlement of the XVIII and XIX centuries, were the excavation covered the full extension of an occupational unit so to investigate spatial horizontal synchronies.

After the ongoing criteria, in this excavation, the differences between subunits were described according to a formalized and standardized questionnaire of the sediment structure and composition that was confirmed by the analysis and quantification of these variables in the sediment composition. The limits of every subsequent subunit were set and registered, taking a general contextual picture of the subunit and some detailed pictures of each square meter before and after the extraction. The subunits were extracted in the inverse order of its deposition. The depth of the surfaces was also measured following a fifty square centimeter grid. The formalized and standardized recording of the sediment structure was completed with the notation of the 3D location and relative situation (geographic orientation, gradient and archaeological/anatomical position) of anthropic residues which means, for instance, all

* The projects developped during these 20 years were supported basically by the Spanish CSIC, Ministerio de Cultura, Ministerio de Educación y Ciencia, Generalitat de Catalunya, EU XII Comission and the argentinean CONICET.
The first project was co-directed three of the autors and by Luis Abel Orquera.
Nuria Juan-Muns, Joan Miquel Lozano, Alfredo Maximiano, Oriol Vicente and Marian Berihuete contributed to this work with some specific analysis.

identifiable and sort of bones fragments longer than 3 cm. We recovered all kind of archaeological evidences on a millimeter screen at the excavation.

Such procedure allowed a preliminary spatial volumetric reconstruction of superposition of the surfaces, the process and the sequence of the site formation.

The volume of each of the 272 subunits isolated on the 32 square meters was measured. Their range varies from 160 liter to just 0,3 liter in a not normal distribution. Their surface, also very variable, ranging from 0,02 to more than 5 square meters. These subunits yielded 22.303 lithic residues and tools, and more than 10.000 bone fragments (excluding fish).

After a statistical test to verify the optimal procedure, fish remains were also sampled because of their little size and the huge amount of them found. The system consists of the homogenization of all the sediment extracted on each subunit and the recovering even of the smallest remains; all this work is carried out by an expert member of the team in laboratory conditions. Fish remains vary from 0 to 228 pieces per liter in the samples of the subunits (Estévez and Vila, 1995).

There was also a significant variation in the composition of the different components of the subunits (fine sediments, granules and small pebbles, mollusks, charcoal, bones and flakes). We recovered 4 liter samples of homogenized sediment from each subunit, in order to estimate the NMI of mollusks for the complete subunit and to see variations in the internal composition of the shell midden. Every component from each sample was weighed and the MNI for the mollusks was calculated.

We submitted the data of 35 randomly chosen subunits of *Túnel VII* site to an inferential statistical test. The application of the Principal Component Analysis to the resulting weighs of the components shows that the relative quantity of every sediment component found in the subunits is not homogeneous and there are important and not random differences among the samples and their components. This test also showed that the most important difference is the relation between the variable *mollusks* and the variable *pebbles*. These differences between the subunits are related to their location in the hut. The 100% of the subunits characterized by high content of pebbles corresponded to the inner part of the hut, whereas the 88,89% of the subunits with higher quantity of mollusks where situated outside of the hut. Half of the mussel hinges yielded the mussel's MNI ranging from 19,7 individuals/dm^3 to 224,2 individuals/dm^3 with a mean of 89,5 individuals/dm^3. Subunit differences estimated by the weigh of the components behaved as those estimated by MNI: subunits with a higher density of mollusks than the mean were located outside the hut, while the subunits inside the perimeter of the hut contained a density of mollusks under the mean. (Verdún, 2005).

Such results implicates the inner-hut maintenance by cleaning outside the shell but letting or intentionally adding granule and small pebbles.

In the excavation of *Túnel VII* site we used Kubiena boxes for micromorphological analysis. Our aim was to verify the hypothesis about the site formation process by the identification of the stratigraphic subdivisions using microscopic evidences of trampling and accumulation and abandon episodes. We toke samples from relevant locations such as shell middens top surfaces, superposed fireplaces... and a couple of big boxes (55*15 cm) including the whole stratigraphy on the place were the main entrance of the huts could be situated.

The overlapping of fireplaces and subunits showed that this site was probably reoccupied in different moments along a maximum period of a hundred years -from the end of the eighteenth until the end of the nineteenth century- as the dendrochronology, the ^{14}C and the European material found in the excavation shows.

After the positive results obtained from the excavation of *Túnel VII* site, the methodology was applied again with important ameliorations and better technological support, on new field research since 1995 at the *Lanashuaia* site, which is situated about 60 kms eastwards. At this moment, we began to explore new techniques for the automation of data recovering and recording using automatic topographic stations, field computers and GIS software.

During the excavations of *Mischiwen III* and *Cabaña Remolino* in 2001 we began to use systematically digital photography with image rectification systems (Vila, Casas, and Vicente, 2006).

This technology made all data processing much faster and easier.

Shamakush VIII: the introduction of automatic recording procedures

In the 2002 fieldwork at *Shamakush VIII* site a more developed and formal recording procedure was put in work. *Shamakush VIII* is a single component site dated 1400 ± 90 (AC 1678) and 1380 ± 115 (AC 1681) ^{14}C uncalibrated BP (Piana y Vázquez 2005).

The field season at this site was designed in order to build a model for a computer-based representation of the shell middens formation processes.

To be able to decompose the archaeological site in their components, we built a geometric representation of measured spatial variables. (Adán, *et al.* 2003 and Barceló, *et al.* 2002) Such variables, as previously used in all of the excavations since the use of the subshellmiddens units, consisted of strata, layers, concentration areas, and ground deformation patterns. They can be described in terms of contact surfaces, which are consequence of a change in the formation process acting on a specific location. In this excavation, the analytical unit used it has always been the upper stratification surface of a deposit. The bottom or basis is, by definition, the upper part of other depositions formed before.

We used a vector based data structure. Each observed discontinuity or contact surface was divided using a precise grid, with nodes at 1 meter. Each square meter was photographed separately, and the resulting digital image was corrected using 5 control points (grid nodes). Point coordinates were measured in the field with a Total Station. The obtained data were sent to the computer where the scale and distortion of the original image were altered in order to coincide exactly with coordinates. Once rectified, the pictures were joined in a mosaic, to obtain a precise image of the observed discontinuity. The border correctly scaled can be transformed into a vector, and the discontinuity or sedimentary area can be described as a polygon made of oriented vectors. The building of a geometrical model of their relationships is, then, a fairly straightforward task.

The photo mosaic becomes a real image map of the ground. It contains landmark data (shape, size, location) and retinal properties (texture, composition). We use GIS software for processing this geometric representation of the observed discontinuities.

A 3D view of each surface contact is easily computed when interpolating measured z values, a volumetric representation of the whole site is also possible. This volume representation of the site sedimentation can be contrasted with a 3D colour view were the represented values can be any other variable different from elevation (another archaeological distinct structure: a qualitative variable, such as the presence or concentration of some material; or a quantitative value, such as the value of some sediment property or any phenomenon that can be measured or found at every location).

The aim of all this development was to simplify and accelerate the field recording as well as the laboratory treatment of all collected data, which is very important because the system of recovering subunits in shell middens requires a lot of time and it produces a huge amount of information which has to be processed and exploited.

DISCUSSION OF THE RESULTS: THE NON RANDOM CHARACTER OF DEPOSITION IN SHELL MIDDENS

Obviously, the structural visualization of a shell midden is not the only purpose of the approaching. After all, an archaeological site is the place where social actions "were" performed. As a result of the social strategies of managing resources and people, physical space is being modified: people excavated pits, they accumulated sediment, they built walls and they broke the original shape, size, and texture of some surface entity. Some times, the nature of past actions did not generate any modification in the main shape, size or texture features of space surface; but it produced a modification of its composition: people accumulated artefacts when they produced garbage and they deleted things when they cleaned the hut.

To understand human activity within the spatial and structure model of our sites in Tierra del Fuego we used ethnoarchaeological knowledge as a reference frame. Yamana constructed perishable huts of about 3,5m of diameter. These huts were occupied only once during a couple of weeks, or repeatedly after abandonment periods. The consumption remains abandoned in the settlements are the main component of the sites.

The excavated sites are mainly circles surrounded by a ring composed by shell middens and other consumption residues originating a simple stratigraphy in some sites or a more complex stratigraphy in others. The archaeological record matches well with the ethnographic descriptions of the shape, size and discontinuous but repeated use of circular huts by yamana people.

In Tierra del Fuego sites, we could relate garbage areas constituted mainly by middens of mussels surrounding other areas that were cleaned.

Given the main human character of the deposited material, we can assume that its deposition structure (form and content) has to do with the organization of human activities. This organization of remains and therefore the human use of space is something that has to be analyzed prior to consider the deposit as a random feature. The possibility of a random nature of deposition has to be demonstrated and explained in any case as a result of entropy introduced by accumulation of random human or taphonomic activity (Mameli, Barceló, Estévez, 2002). If we assume from the beginning a hazardous or non-structured distribution and if we use a consequent sample technique, then we can hardly discover the hidden organization nor can we demonstrate the actual performance of the sample we have token. This concept led the development of the subshellmidden excavation from its very beginning (Orquera & Piana 1992). Instead, we better develop the search for the finest significant differences trying to explain them.

As an example we can show some results and simulations obtained on our sites.

Explaining the formation process: micromorphological studies

As we have explained before, the identification of subunits followed a visual recognition aided by a qualitative and quantitative description according to standardized variables and tested by the detection of stratigraphic discontinuity surfaces. This was tested when the measurements and the pictures were already taken. Notwithstanding that micromorphology has been profusely used in archaeological studies (Solé and Vila, 1991; see also French 2003 for a review), its application to understand midden structure and stratigraphy has been neglected (an exception is the work of Simpson and Barrett 1996). Micromorphology samples were collected during the several field season (Taulé 1995) as an integral part of the applied methodology for providing information. That was done through the analysis of the pedofeatures, on the spatial variability of the deposits, on the depositional histories (e.g. the episodes of formation) of the sites and on the stratigraphic superpositions that constructs the shell middens.

One of the most important results is the presence of a number of micromorphological features that point to the occurrence of post-depositional processes which permit to identify units of deposition. These are the related to near-surface pedogenesis features (e.g. textural and excremental) that suggest a short-term exposure of the shell midden units. Most of the organic material shows a good degree of decomposition suggesting high biological activity. Exposed surfaces are also often characterized by a compacted level of *in situ* broken shells originated from trampling (Figure 15.1). Besides, probably secondary deposits of broken shells are observable, which would result from rearrangements of the deposits during or prior an episode of occupation. On the basis of these preliminary observations, it has been possible to highlight episodes of depositions that represent the change of depositional focus and structure reorganization that happened during the shell midden/hut life. The micromorphological studies helped to the understanding of the actual formation process and therefore to test the validity of the field identification of subdivisions. The big Kubiena boxes will allow comparing the record of the deposition sequence gathered during the excavation with the formation processes revealed by the microscopic observation. The comparison of this evidences as well as the information obtained from other applied methodologies has helped in elucidating the formation processes of these peculiar sites and has shown the importance of micromorphology for understanding shell middens structure.

Establishing the stratigraphic sequence:

The representation of the superposition of subunits and its correlation follows a system like the Harris Matrix that is elaborated from the superposition of the maps of every subunit (Estévez and Vila, 2000). The superposition of the marks left by the fireplaces on the abandoned surfaces during the successive occupations (a sequence of colors – from black to light yellow- result of the temperature and fire action on the soil) allows the establishment of a sequence of occupations. The subunits can be related to the different episodes of occupation by their overlapping. In *Túnel VII* site there were at least ten discrete re-occupations (Estévez and Vila 2006), whereas in *Lanashuaia* (Piana, Estévez, Vila, 2000) there were no more than two occupation episodes, which in any case are difficult to isolate (figure 15.2).

Some of the subunits show an ambiguous position in the process of "rebuilding" of the stratigraphy because they did not overlap with the subunits that allow the segmentation of the sequence of occupation episodes. And, of course, despite the fast sedimentation rate of shell middens, we don't think it could exactly be a Pompeiic process. But in such cases the record of the exact position of items and the fifty centimeters grid allow the use of refitting bones and lithics to help in this stratigraphic analysis and to solve such problems of alternative stratigraphic positions of some subunits and of taphonomic movements.

In *Túnel VII* refitting of bones was helpful for eliminating some ambiguities in the correlation of subunits in the time sequence and to look to the synchronic dynamics. The bone refittings do not have all the same discriminatory value for this proposal. We classify them in different kinds and probabilities: repairs (re-assembling of ancient broken bones), rearticulations of exact matching bones and hypothesis (sure, probable and possible) about reassembling of individual skeletons. We have to consider also the state of conservation of the bones and the marks on its surface (trampling, polishing, rolling, weathering), as well as its place (on the bottom, body or top) in the subunit and the way they lie (horizontal, vertical, following the shell orientations and so on).

In the 32 square meters of *Túnel VII* site we have measured 10.500 items exactly and the rest of items in a 50 cm grid. From a total of 10.476 faunal pieces we found 1.555 refittings. We established 76 repairs for bird bones, which was really a hard task. We also established 443 refittings for sea lions and 13 for guanaco. Forty four parts of different sea lion skeletons could be rearticulated. But the high quantity of such refittings allows filtering the most significant ones in a qualitative and a statistical way: 25% of these refittings, link different subunits. Looking closer to them we can link subunits that were first not stratigraphically related. For instance we could follow some dynamics of deposition related to people's activities and space management: a part of a skeleton lies on the bottom of a midden subunit; it was covered very quickly by garbage. Another piece was incorporated in the body of a shell midden subunit produced by wiping away the garbage. These elements of the skeleton are not in the anatomic connecting position and show a slight trampling on the bone surface. Another piece has been deposited on the top of a third shell mound. This last element is characterized by a higher degree of trampling and weathering.

The refitting of stones artifacts has also a different weight depending on the type of residues involved.

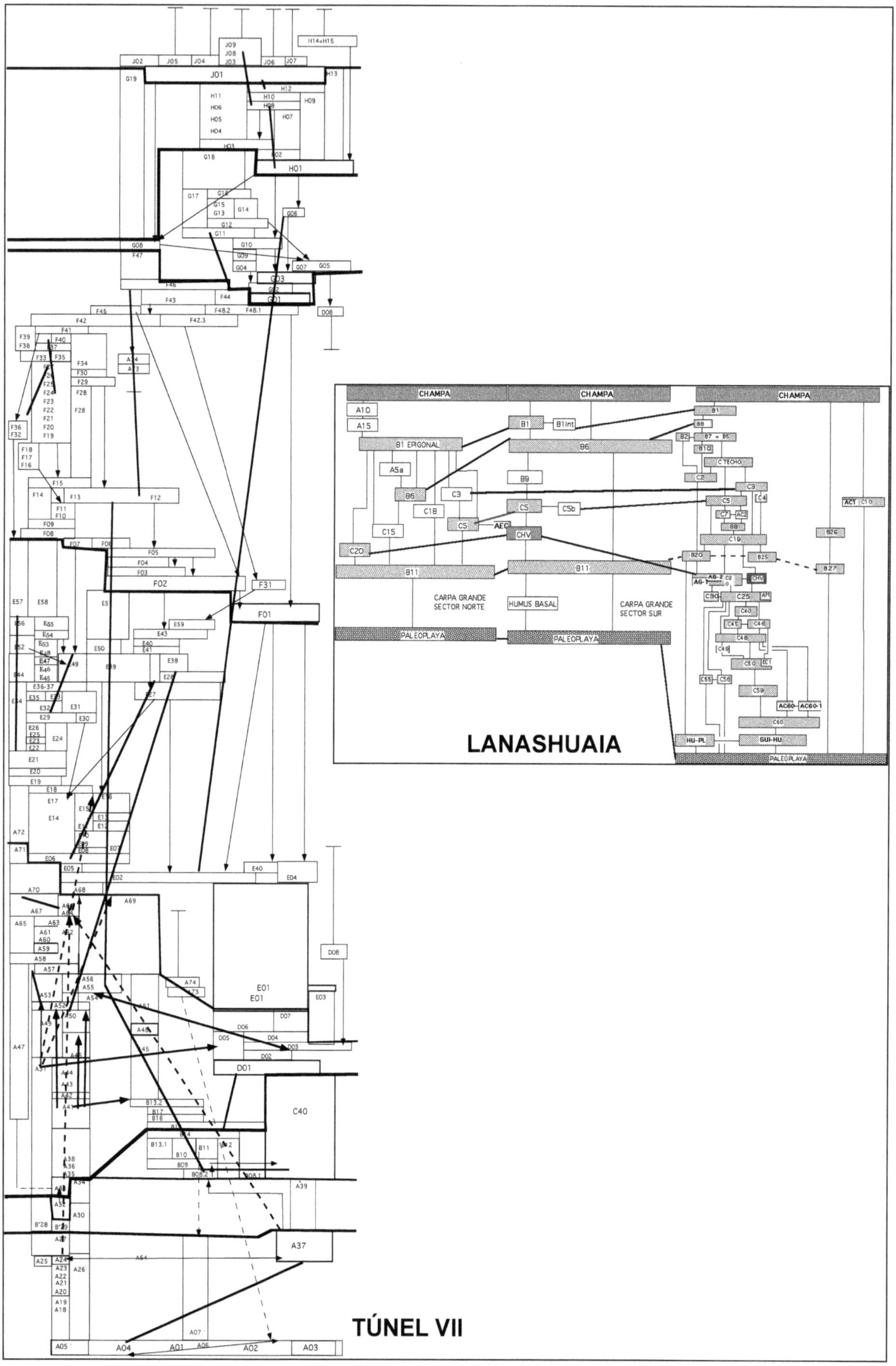

Fig. 15.1. Stratigraphic matrix of the sets *Túnel VII* and *Lanashuaia*

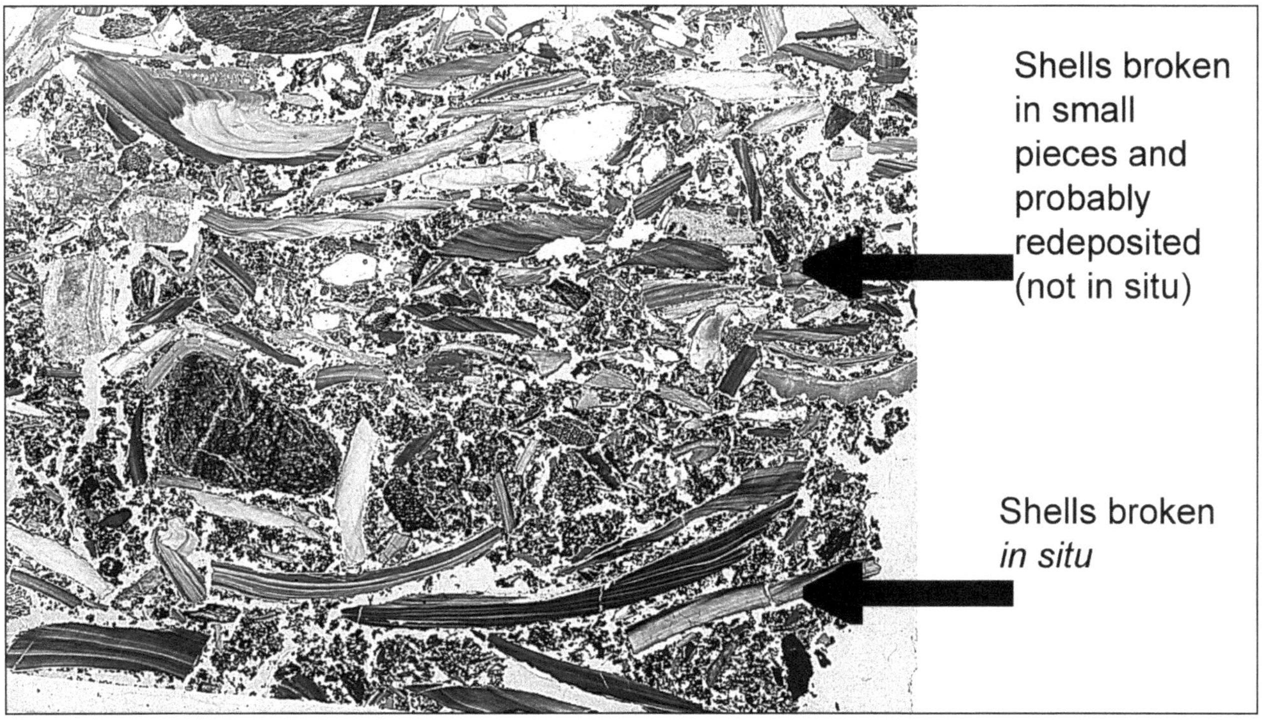

Fig. 15.2. Micromorphology section of a sample in one of the Kubiena boxes of T*únel VII* showing episodes of reworked an in situ broken shells

We establish three types of categories concerning to: refitting core to flake, flake to flake and on the same flake.

Synchronic variation of remains distribution in shell middens: the need for an extensive excavation approach:

The simulation of sampling procedures showed that we would not get an adequate picture of the global content of the site with samples of a half square meter of the whole stratigraphic column, nor with samples of a square meter.

In *Túnel VII* the distribution of all bone remains of the vertebrate taxa in distinct squares meters offers significant differences.

Tab. 15.1.

	square 1/2	square 5/3	square 10/4	square 15/5
Birds	64%	54%	53%	35%
Whales	4%	2%	5%	26%
Guanaco	1%	2%	1%	1%
Sealion	29%	40%	35%	32%
Non det.	3%	2%	5%	7%

The faunal samples deposited in different spots over an occupation episode or over the whole sequence vary significantly from the center to the peripheries of the occupation area.

But even the frequencies by half square meters inside the same square can vary significantly.

Tab. 15.2.

	square 5-3	square 5-4	square 6-3	square 6-4
Birds	68%	40%	47%	67%
Whales	1%	1%	3%	4%
Guanaco	3%	1%	3%	1%
Sealion	28%	53%	45%	28%
Non det.	0%	5%	3%	0%

In order to state horizontal variability in deposition, which we consider in our case an outcome mainly of anthropic processes, phytolith analysis was carried out on 12 samples from one of the depositional subunits (B355). This subunit is located inside of the shell ring, which represents the inside of a hut. The phytolith assemblages clearly show meaningful differences in their composition and densities, which depend on the location of the samples. One of the samples has a very high phytolith density, which means a higher input of plant material. This sample contains jigsaw-type phytoliths from *Blechnum penna-marina* (Poiret) Kuhn, a fern of Tierra del Fuego that typically grows in close forests. This implies the gathering of the species outside and a special accumulation of its debris close to this spot, which is located close to the limit of the hut. This special accumulation coincides with the use of these plants for conditioning and covering the hut walls and roof as described in the ethnographic record.

Tab. 15.3.

Episodes	Embotrium	Drymis	Ribes	Chilothrium	Berberis	Maytenus
A	64%	25%	7%	3%	0%	0%
B	35%	42%	14%	8%	0%	1%
C	30%	36%	22%	9%	1%	0%
D	44%	31%	16%	7%	2%	0%
E	40%	41%	13%	4%	1%	0%
F	44%	37%	13%	5%	1%	0%
G	32%	44%	18%	5%	1%	1%
H	35%	39%	16%	7%	1%	1%
J	52%	29%	12%	4%	2%	1%
TOTAL	4521	4035	1693	647	119	48

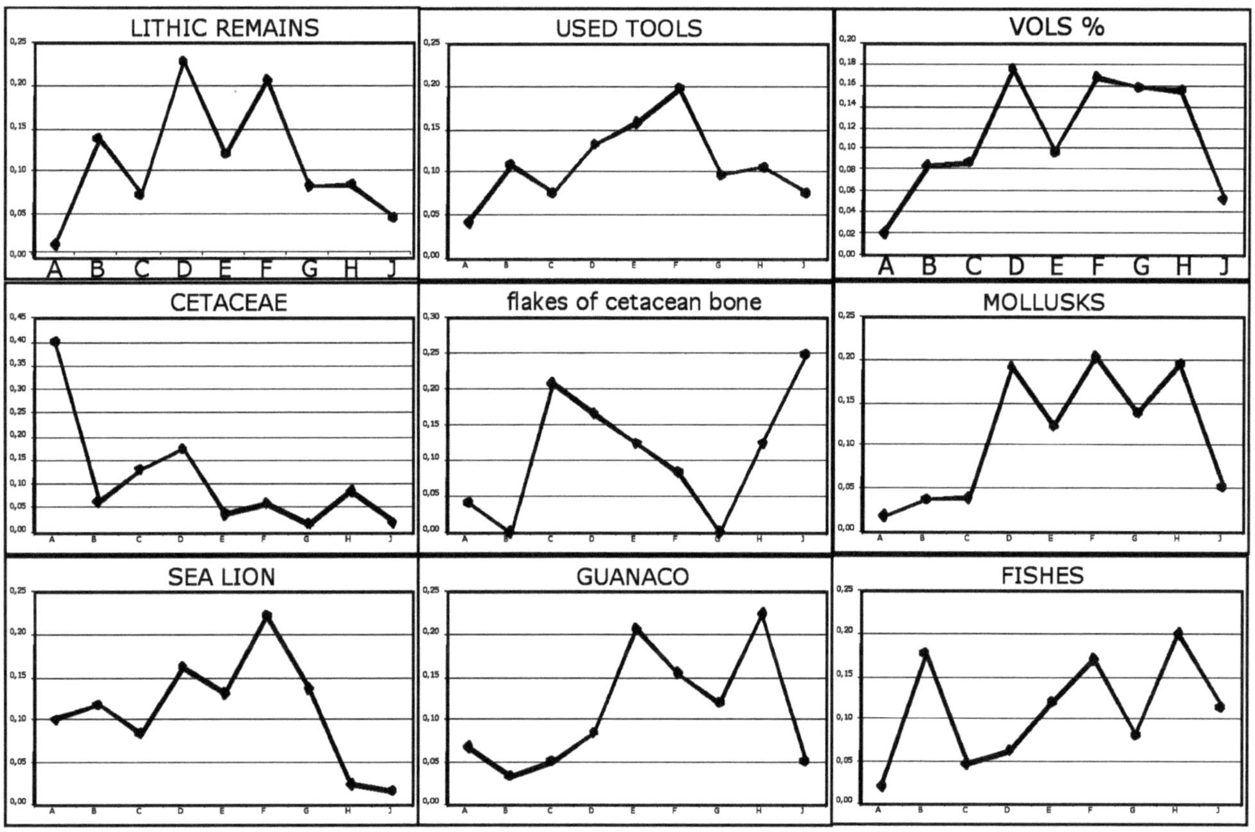

Fig. 15.3. Relative frequencies oscillations in several main items along the sequence of occupation episodes in *Túnel VII* site

The distribution of stone tools and rejects of the stone tool production is also not even but very meaningful (Clemente and Estévez, i.p.) showing specific tendencies of accumulation.

The diachronic variability: the need of dissecting thin episodes

In the sequence of *Túnel VII* occupations, that spanned just over a century we stated a statistically significant variation in all types of consumption goods.

The relative composition of firewood along the sequence of episodes shows significant differences (Piqué, 1999) as shown in the table 15.3.

Actually all the categories vary along the sequence. Every occupation episode shows a different composition in the archaeological items recovered. They do not co-vary along the sequence of episodes, nor are they strictly caused by a seasonal organization of the activities (figure 15.3). That means that every single episode does not represent a meaningful sample of the global strategies over a century. A global picture of the yamana exploitation and subsistence strategies does not emerge with a simple sum or a mean of the activities fulfilled. Their variability and flexibility would be missed if we could not isolate the samples of every episode (Estévez and Vila, 2006). This is why the dissecting during the excavation of the minimal occupation episode is justified.

DISCUSSION: EXPLAINING THE FORMATION PROCESS

In fact, refitting as well as the other spatial sampling techniques and stratigraphic isolating of subunits, and the establishment of a detailed sequence of occupations was not only designed to fix stratigraphic links and to control post-deposition processes but also to accomplish the identification of activities and reject areas, intra site circulation of products and articulation of working activities: processes of flaking, re-sharpening and recycling tools and other items, butchering, dismembering, cooking and consuming food.

Plotting the different residues types allows depicting the dynamics of the work and the distribution of the consumed goods. It makes evident the wall effects helping to understand the organization of space (Estévez and Vila, 2006, Clemente and Estévez, i.p.).

The potential variability between stratigraphic subunits (in a vertical diachronic or horizontal synchronic sense) depends on the organization of activities of production, distribution, consumption and deposition as well as on the inequality of consumption and on the recurrence, normalization, regulation and fixing of habits of discarding residues and of arrangement of space (Zurro, *et al.* 2006).

We can verify the recurrences and the variability in the strategies of subsistence, in the organization of space and in the distribution of real value (the labor effort invested in the production of the item) or the consumption value of the products of labor (Barceló *et al.*, 2006).

In sum: a shell midden is not necessarily a homogeneous block of sediments. The variability inside can be very significant of the economic and social system that produced it and a sample of homogenized sediment did not represent but obscure this information. It is a good goal to try and to improve finer dissecting of shell middens and other sites.

References

ADAN, M.; BARCELO, J.A.; PIJOAN-LOPEZ, J.; PIQUE, R.; TOSELLI, A. (2003) Spatial Statistics in Archaeological Texture Analysis. In The Digital Heritage of Archaeology. Computer Applications and Quantitative methods in Archaeology. M. Doerr and A. Sarris (ed) Hellenic Ministry of Culture. Archive of Monuments and Publications.

BARCELO, J.A.; BRIZ, I.; CLEMENTE, I.; ESTEVEZ, J.; MAMELI, L.; MAXIMIANO, A.; MORENO, F.; PIJOAN, J.; PIQUÉ, R.; TERRADAS, X.; TOSELLI, A.; VERDÚN, E.; VILA, A. and ZURRO, D. (2006) Análisis etnoarqueológico del volr social del producto en sociedades cazadoras recolectoras. In: *Etnoarqueología de la Prehsitoria más allá de la analogía*. Departament d'Arqueologia i Antropologia IMF-CSIC. Barcelona. p.189-208.

BARCELO, J.A. de CASTRO, O., TRAVET, D., VICENTE, O. (2002) Towards a 3D representation of Archaeological Layers. Archeologie und Computer vol. 6. Wien, Phoibos Verlag.

CLEMENTE, I. and ESTÉVEZ, J. (2006) Space Management During the Firs Occupations of the Site Túnel VII (Tierra del Fuego, Argentina). Conference: The Archaeology of Household. Barcelona. (i.p.).

ESTEVEZ, J. and VILA, A. (eds.) (1995) Encuentros en los conchales fueguinos. Treballs d'Etnoarqueologia, 1, Barcelona: CSIC-UAB.

ESTEVEZ, J. and VILA, A. (2006) Variability in the lithic and faunal record during ten reocupations of a XIX century Yamana hut. Journal of Anthropological Archaeology, vol 25(4), p. 408-423.

ESTEVEZ, J. and.VILA, A. (2000) Estratigrafias en contexto. Krei n°5 , p. 29-61.

ESTEVEZ, J. and VILA, A. (1995) Etnoarqueología: El nombre de la cosa. Treballs d'Etnoarqueologia, 1, p.17-23 Barcelona:CSIC-UAB.

FRENCH,C. (2003) Geoarchaeology in action: Studies in soil micromorphology and landscape evolution. London: Routledge.

MAMELI, L. BARCELO, J.A.; ESTÉVEZ, J. (2002) The Statistics of Archeological Deformation Process. A zooarchaeological experiment. Archaeological Informatics: Pushing the Envelope. G. Burenhult (ed) p. 221-230 (BAR Int. Series 1016).

ORQUERA, L.A. and PIANA, E.L. (1988) Human littoral adaptation in the Beagle Channel region: the maximum possible age. Quaternary of South America and Antartic Peninsula, 5, p.133-165.

ORQUERA, L.A. and PIANA, E.L. (1992). Un paso hacia la resolución del Palimpsesto. Análisis espacial en la arqueología patagónica. In L.A. Borrero and J.L. Lanata (compil.). Buenos Aires. Ed. Ayllu. p.21-52.

ORQUERA, L.A. and PIANA, E.L. (2000) Composición de conchales en la costa del canal Beagle (Tierra del Fuego, República Argentina) Primera Parte. Relaciones de la Sociedad Argentina de Antropología Tomo XXV, p.249-274.

ORQUERA, L.A. and PIANA, E.L. (2001) Composición de conchales de la costa norte del Canal Beagle (Tierra del Fuego, República Argentina) Segunda

Parte. Relaciones de la Sociedad Argentina de Antropología Tomo XXVI, p.345-368.

ORQUERA, L.A.; SALA, A.E.; PIANA, E.L.; TAPIA, A.H. (1978) Lancha Packewaia: arqueología de los canales fueguinos. Buenos Aires: Huemul.

ORQUERA, L.A.; PIANA, E.L.; ÁLVAREZ, M.R.; FIORE, D.; VÁZQUEZ, M.M.; ZANGRANDO, A.F.J., TESSONE, A.Y., TIVOLI, A.M. (2006). El Proyecto Arqueológico Canal Beagle. *Arqueología de la Costa Patagónica. Perspectivas para la Conservación* (compil.: I. Cruz, M.S. Caracotche, M. X. Senatore y B. Ladrón de Guevara). Cap- 16:288. Río Gallegos.

PIANA, E.L. y VÁZQUEZ, M.M. (2005). El Sitio Shamakush VIII: puntualizaciones sobre el uso de recursos y la gestión del asentamiento en el canal Beagle. In Actas *XV Congreso Nacional de Arqueología Argentina*. (i.p.).

PIANA, E.; ESTEVEZ, J.; VILA, A. (2000) Lanashuaia:un sitio de canoeros del siglo pasado en la costa norte del canal Beagle. Desde el País de los Gigantes. Tomo II, p. 455-469.

PIQUÉ, R. (1999) Producción y uso del combustible vegetal: una evaluación arqueológica. Treballs d'Etnoarqueologia, 3. Madrid: CSIC.

SIMPSON, I.A. and BARRETT, J.H. (1996) Interpretation of midden formation processes at Robert's Haven, Caithness, Scotland using thin section micromorphology. Journal of Archaeological Science 23, p.543-556.

SOLE, A. and VILA, A. (1991) La micromorfología de suelos aplicada a la arqueología: dos casos a modo de ejemplo, el Cingle Vermell y Mediona I. In Recent Developments in Western Mediterranean Prehistory: Archaeological Techniques, Technology and Theory. Oxford B.A.R., n°574 vol.I.

TAULÉ, A. (1995) Primeros trabajos micromorfológicos en concheros de la costa del Canal Beagle. Treballs d'Etnoarqueologia, 1, p.113-126 Barcelona.CSIC-UAB.

VERDÚN CASTELLÓ, E. (2005) Aproximació a l'estudi dels conquillers a través de la malacofauna. Trabajo de Investigación de Doctorado, Departament de Prehistòria, UAB. Barcelona.

VILA, A. (2004) Proyectos etnoarqueológicos en Tierra del Fuego (Argentina). Bienes culturales, n°3 (Excavaciones arqueológicas en el exterior), p. 193-200.

VILA, A. (2006) Propuesta de evaluación de la metodología arqueológica. Treballs d'Etnoarqueologia, 6, p. 61-76.

VILA, A. and WÜNSCH, G. (1990) Un pequeño paso antes del gran salto: buscando como preguntar (o investigación teórico-Metodológica en Tierra del Fuego) Xàbiga, 6, p. 19-31.

VILA, A.; CASAS, A.; VICENTE, O. (2006) Mischiuen III, un contexto funerario singular en el Canal Beagle (Tierra del Fuego) Revista Española de Antropología Americana n°36-1: 47-61

WASELKOV, G.A. (1987) Shellfish gathering and shell midden archaeology. Advances in archaeological method and theory, vol. 10, p.93-210.

ZURRO, D., BRIZ, I., ÁLVAREZ, M. and VILA, A. (2006) The integrity of social space: Hunter-gatherers at the end of the world. Communication. Conference: The Archaeology of Household. Barcelona. (i.p.).

SHELLMIDDENS OF THE ATLANTIC COAST OF NICARAGUA: SOMETHING MORE THAN MOUNDS

Ignacio CLEMENTE CONTE

Laboratorio de Tecnología Prehistórica, Departamento de Arqueología y Antropología, Institución Milá y Fontanals, Consejo Superior de Investigaciones Científicas (CSIC), Barcelona. C.e.: ignacio@bicat.csic.es

Ermengol GASSIOT BALLBÈ

Departamento de Prehistoria, Universidad Autónoma de Barcelona (UAB), Unidad Asociada al CSIC. C.e.: ermengol.gassiot@uab.es

Leonardo LECHADO RÍOS

Centro Arqueológico de Documentación e Investigación (CADI), Universidad Nacional Autónoma de Nicaragua (UNAN- Managua). C.e.: lechadorios@hotmail.com

Abstract: *Usually excavations of shellmiddens in the Middle America have been test pits of reduced area in order to know the characteristics of their internal stratigraphy and to recover both sediment and artefactual samples. Nevertheless recent extensive excavations of shellmiddens in the Americas proved their utility to understand these kinds of sites as by-products of past social life. This paper discusses the results of the extensive excavation of a shellmidden in the Karoline site, in the central Caribbean coast of Nicaragua. The results of the fieldwork and the later analyses of the archaeological remains allow us to know both aspects concerning the social life in this old village (occupied between 450 calBC until 350calAD), the use of the domestic space inside and surrounding a hut and the process of formation of the shellmidden close to this habitation area. More than a simple midden dump, the shellmound is also a place where some activities were carried out, e.g., intensive cocking. Moreover, the detailed view of the internal stratification shows a large process of sedimentation of the shellmidden, with a constant reworking of the internal work areas.*

Keywords: *Shellmiddens, Archaeology, Caribbean Coast, Nicaragua*

Résumé: *Normalement lors des fouilles archéologiques dans des amas coquilliers d'Amérique Centrale n'ont été prélevés que des échantillons de petites tailles afin de connaître les caractéristiques de leur stratification interne, et de récupérer des sédiments ainsi que des artefacts. Néanmoins, d'autres fouilles plus récentes d'amas coquilliers en Amérique ont démontré leur utilité pour comprendre ces sites en tant que produits de l'activité sociale du passé. Cet article expose les résultats d'une fouille en extension réalisée dans le site de Karoline, sur la côte centrale de la mer des Caraïbes du Nicaragua. Le résultat des travaux sur le terrain et les analyses postérieures des matériaux archéologiques nous a permis de connaître des aspects concernant la vie sociale de cette ancienne petite bourgade (occupée entre 450 Cal. BC et 350 Cal. AD), l'usage domestique de l'espace à l'intérieur et dans les alentours proches d'une unité d'habitat (cabane/maison) et le processus de formation de cet amas en relation avec cette unité. Plus qu'un simple monticule de débris, l'amas coquillier est aussi un lieu (ou aire) où certains travaux ont été accomplis comme par exemple la préparation d'aliments. De plus, l'examen détallé de la stratification interne dévoile un vaste processus de sédimentation de l'amas coquillier, avec une modification constante des zones internes de travail.*

Mots clés: *Amas à coquillage, Archéologie, Côte du Caraïbe, Nicaragua*

INTRODUCTION

The Atlantic coast of Nicaragua is a relatively flat area of tertiary and quaternary alluvial deposits with numerous lagoons and mangrove swamps. It is currently populated by an amalgam of different ethnic groups, the *Misquito* and *Rama* descendants of indigenous groups native to the Caribbean coast of Nicaragua, the *mestizo* population proceeding from the west and centre of the country, the *Garifona*, descendants of African slaves and workers, and the *Creolle*, a mixture of the former and the latter. Today, a large percentage of this population works on the large plantations (sugar cane, palm, bananas, etc.) while others are small peasants who mainly cultivate maize, cassava, pumpkins, and frijoles and keep some domestic animals, complementing their diet through the hunting of wild animals (peccary, deer, iguana, etc.). The coastal populations still exploit the aquatic recourses intensely and they practise fishing, turtle rearing and, naturally, seafood gathering (Nietschman 1973). Even today, one can see in various communities how shellmiddens (of clams and oysters) are formed at one end of the wooden dwellings, which are usually also raised above the ground to protect them from the damp caused by the intense rains that are frequent during most of the year in this tropical humid area (Figure 16.1).

Most of the archaeological works carried out in the Atlantic coast of Nicaragua have been directly related to excavations of sites with "shellmiddens" (Magnus 1974-1978, Matilló 1993) and, until our works in the zone, there was hardly any radiocarbon dating. These archaeological actions were generally test pits of two square metres, excavated by artificial arbitray levels. These tests were conceived fundamentally as a means of recuperating archaeological artefacts, mainly sherds, which allowed us to make a chrono-cultural sequence that is applicable to this geographic zone.

Fig. 16.1. Modern Shellmidden in the Neighbourhoods of Bluefields

With this degree of archaeological knowledge for this zone of the Caribbean, and with the supposition that these coasts were inhabited by small groups who moved frequently and who hunted, gathered, fished and practised a probably very incipient type of agriculture, as documented in the ethnohistorical sources (Roach 1991, Romero 1995, Squier 1891), we began a series of research and co-operation projects between the *Universidad Autónoma de Barcelona* (UAB) and the *Universidad Nacional Autónoma de Nicaragua* (UNAN-Managua) in the second half of the 1990s, with the participation of the *Consejo Superior de Investigaciones Científicas* (CSIC) since 2002, from the first archaeological excavation seassons. The explorations carried out (1998-2000) along the coast between Bluefields Bay and the Pearl Lagoon, allowed us to document at least 22 archaeological sites with a total of more than 80 shell mounds or shellmiddens.

THE PREHISTORIC AND HISTORIC SHELLMIDDENS BETWEEN BLUEFIELDS AND PEARL LAGOON: GENERAL CHARACTERISTICS

The radiocarbon dating of these shellmiddens offered us a range of dates between 1410-1270 calBC (Coconut's Beach) and 782-1021 calAD (Cukra Point) (Gassiot 2005, Gassiot and Palomar 2006, Gassiot *et al.* 2003). The species of molluscs most represented in these shell-middens are: clams (*Polymesoda solida*), "*ají*" (*Donax denticulatus* and *Donax estriatus*) and mangrove cupped oysters (*Crassostrea rizophorae*). The clams and oysters are bivalves and they live in the semi freshwaters of the various bays and lagoons of this geographic area, while the *Donax* live in more salubrious environments, forming big colonies or banks in the intertidal zone of open sea beaches during the less rainy periods of the year, when the supply of fresh water from the rivers falls.

The manner of gathering these species is also different. On the one hand, the clams are collected by hand after they have been located in the shallow and soft beds of the lagoons using the feet. This task is currently carried out by boys and girls from the village. The clams are consumed locally and only in the coastal villages. On the other hand, oyster collecting occurs on a fairly more intense basis in the Bluefields Bay area, where the activity employs a large percentage of the members of some Rama communities. The oysters are gathered manually in the banks where this species appears, also in shallow waters and generally on hard substrate. Unlike the case with clams, the oysters are sold commercially in addition to domestic consumption. Finally, *Donax* sp. is gathered with the help of tools (rakes, spades, baskets...) and no type of selection is made on the basis of size.

Practically all of the almost 90 shellmiddens localised within the framework of the current research on the Atlantic Coast of Nicaragua are formed by one of this kind of bivalves, which represent more than 80% of the whole (Gassiot 2005). In most cases this type is *Polymesoda solida*. Only a group of shellmiddens in the part of Bluefields Bay contains a dominant amount of *Crassostrea rizophorae* and only the shell mounds of the Karoline site are formed by an accumulation of *Donax* sp. The taxonomic distribution of the shellmiddens seems to respond more to their geographic position than to chronological criteria, as evidenced by the synchrony between settlements such as Karoline, with a succession of occupations dated between 450 calBC and 400 calAD and others where *Donax* sp. is not present in shellmiddens formed exclusively by *Polymesoda solida*, as occurs at the Sitetaia and Brown Bank sites. In most cases these are located in points relatively high above the current bay coast or the paleocoast corresponding to Holocene transgressions. The exception to this are the shellmiddens of Coconut's Beach and Long Mangrove, whose chronologies are the oldest of the sequence documented up to now and which are to be found in very low areas. The sediment which completes the matrix of these shellmiddens is composed of very organic limes which seem to have been formed within a flooding context. We can add the complete absence of archaeological artefacts to this fact. A hypothetical explanation for the characteristics of these two shellmiddens is that they correspond to points for the processing of the shells once they have been gathered, to extract their meat and move it to other places, further inland, where it would have been consumed.

ARCHAEOLOGICAL EXCAVATIONS IN SHELLMIDDEN NUMBER 4 (KH-4) OF THE KAROLINE SITE (KUKRA HILL, RASS)

In 2002 we began the first excavation session in shellmidden number 4 of the Karoline site. This site is characterised by the 13 shell mounds that have been identified up to now. Another mound, larger in size which is situated in the centre of the dispersion which defines the shellmiddens and shows no remains of valves or other food waste. We do not yet know its specific functionality or its absolute chronology. We have only been able to confirm that it was built with the contribution of tonnes of earth and that they used stones for the construction of walls. The ceramic fragments, as with the lithic remains recuperated in this mound, are of the same raw materials and shapes similar to the rest of the Karoline site, features which allow us to raise the hypothesis that this constructive structure was used synchronically in the formation of the different shellmiddens.

The distribution of the Karoline shellmiddens suggests that each of them may correspond to dumping places for the waste of a habitation unit and that, just as happens today, the "house" or hut may be situated in the space next to it. On the basis of this information, we decided to excavate one of them extensively to obtain information about the occupied space and the activities carried out in it as well. For this purpose, the open excavation area (50 m^2) covers both the part of the shellmidden to the north (15 m^2) and the adjoining space (Fig. 16.2).

In these first excavation campaigns we have tried to apply an excavation technique by layers and sublayers, just as we have done in shellmiddens of Tierra del Fuego (Orquera 1996, Orquera and Piana 1992 and 1996; Estévez and Vila 2000), to try to control the different periods of deposition and to relate them to the different levels identified outside the shellmidden. We believe that although his excavation method is slower than excavating in arbitrary level tests, it is more beneficial in terms of the information that it provides as it has allowed us to determine a series of structures that we would not have been able to identify using another excavation procedure. Accordingly, now we can define the surfaces of the shellmiddens not only as a "rubbish dump" or place of accumulation for garbage or productive waste (mollusc valves, remains of fauna eaten, ceramic sherds, lithic remains etc.), but also as an area of the settlement where different structures were built and where specific productive activities were carried out.

The first thing that attracts our attention in the southern excavated part of the shellmidden is that the contact with the adjacent zone is not natural but rather it presents a series of excavated parts, mostly rectangular in shape and clearly anthropic in origin. In these structures we can find clear signs that they housed hearths and in some of them it is even possible to observe different moments of combustion. These combustion structures are clearly related to a series of "post holes". These may have been used for the construction of either windshields or "defensive fire structures" or else structures for the cooking or smoking of food. The large amount of fish recuperated in the shellmidden allows us to suppose an important exploitation of these aquatic resources. Smoking, especially in these hot and humid zones, is a method that is frequently used for the conservation of fish, and even today we have been able to observe its application. The collocation of the fires directly in the zone where the shell waste accumulates could also have an environmental or climatic explanation. These sediments are very porous and so the heavy and abundant rains that fall in this area of the Caribbean are quickly filtered and the ground is soon read again to revive or re-light a fire on it. On the other hand, the muddy or clayey sediments of the rest of the site take much longer to dry and are less practical for the positioning of the fires.

On the surface of the shellmidden, especially in its eastern part, we have localised two large "stains" with charcoals and valves corresponding to periods when the fireplaces were cleaned, located principally in the west and central area of the excavated part of the shellmidden (Fig. 16.3 nº 4). This layout of the dwellings allows us to hypothesise, at least in the last periods of occupation of habitation unit number 4 at Karoline, that while the western zone of the shellmidden was used for situating different combustion structures, the eastern part was where the different types of waste, both food waste and production and/or maintenance waste were deposited.

As we have mentioned above, in the surface of the shellmidden we have localised the presence of numerous post holes (*cf.* Figs. 16.2 and 16.3). These are of different sizes, in terms of both width and depth, and different in terms of their sedimentary composition. With regard to the classification of the post holes according to the sedimentation with which they were filled, we have been able to differentiate three types. On the one hand holes

Fig. 16.2. Map of KH-4 showing the locale of the shellmidden in the north

filled mainly with *Donax* valves; others whose principal component is humus and others filled with earth with valve fragments. On two occasions we have observed in the bottom of the same the presence of stones or ceramic fragments, in a vertical position, which acted as wedges. This sedimentary difference of the filling of these holes suggests that in some cases the position of the structures could have changed during their use. In these cases, when the post is removed it leaves a hollow, which is filled immediately by the sediment which continues to deposit itself throughout the shellmidden formation process. This has been appreciated by the fact that most of the valves that fill that space have been hurled and remain in a vertical position rather than a flat one which is their natural sedimentation. On the other hand the post holes filled with humus seem to be the result of the decomposition of the actual wood *in situ*. If this is the case, we can suppose they may be components of the last structures that still remained when the site was abandoned for the last time. As the surface of the shellmidden is cleaned during the excavation process, we can clearly identify the position of these post holes by confirming how, for example, valves in a vertical position form a small circle; or as a large circular stain of darker earth is clearly differentiated on the surface of the more whitish shellmidden (Fig. 16.3 n° 1-3).

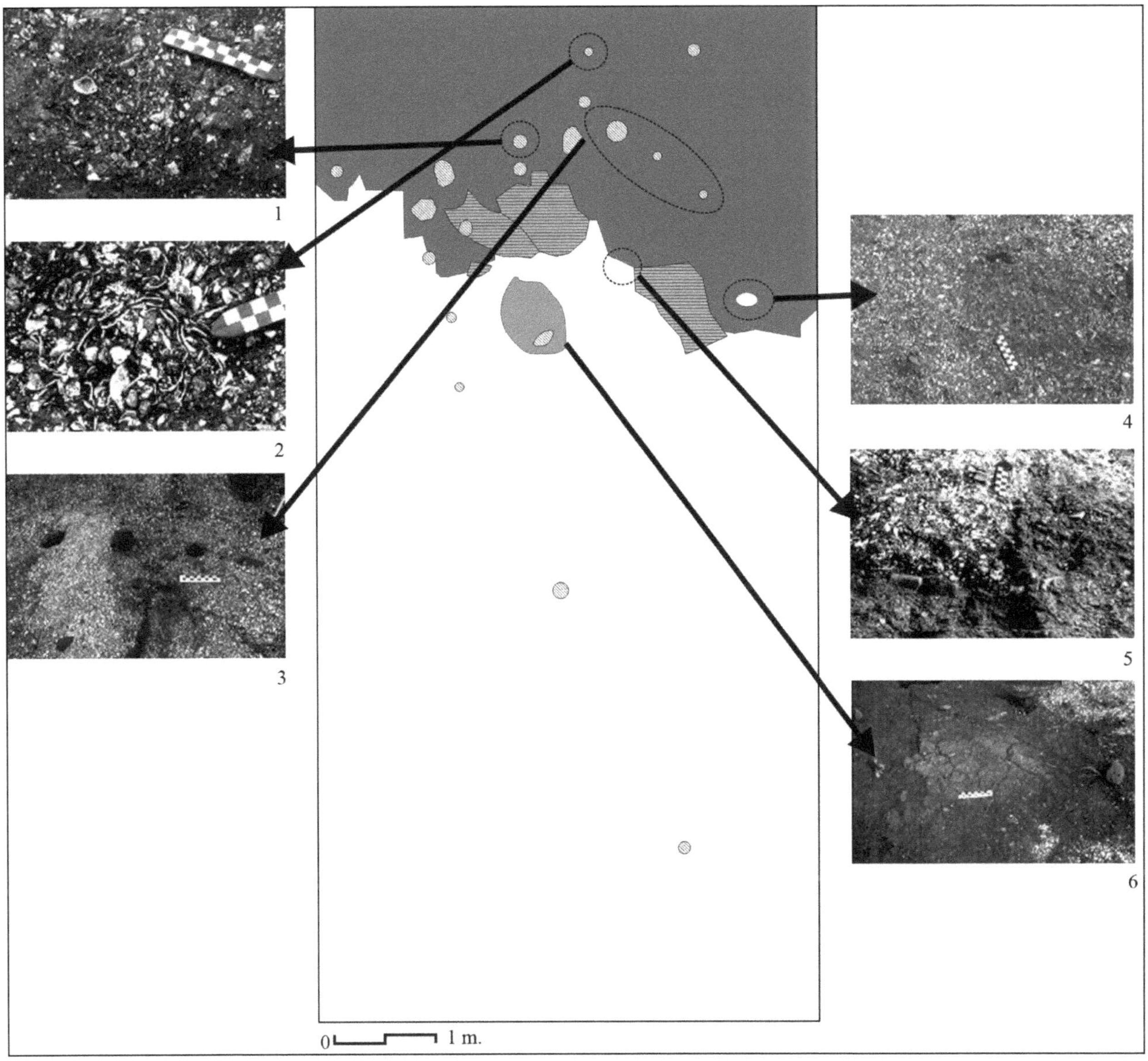

Fig. 16.3. Details of the structures discovered at the KH-4 site

The detailed and meticulous excavation of the contact between the shellmidden and its immediate periphery has also allowed us to observe how at the latter stage of the occupation the sedimentation coming from the south superimposed a part of the sloped shellmidden – in a N-S fall (Fig. 16.2 nº 5). This fact allows us to suppose that the part of the shellmidden is prior to the moment of the formation of that ground and accordingly we can try to relate the external occupation levels with part of the actual shellmidden.

As far as the ceramic materials recuperated in the zone near the shellmidden are concerned, it is interesting to note that in the nearest surrounding area the recuperated vessels are mostly suitable for cooking as their bases show traces of having been in contact with fire. On the other hand, in the area furthest from the shellmidden the ceramic forms recuperated correspond to dishware (plates, glasses, etc.).

Likewise, in the central part of the excavation and in an area close to the shellmidden, a large spot of mud (clay) with a specific coloration and compression from having undergone a thermal alteration has been documented. This spot extends for a little more than one a square metre and its surface does not show any other stigmas – burnt materials, coals, etc.- resembling a combustion structure. However, once it had been excavated, a bed of valves burnt with coals appeared at its base. The alteration of that clay might be due to that lower fire and that it is the remains of the fallen walls of an oven (Fig. 16.3 nº 6).

The vast majority of the lithic remains recuperated in this habitation unit were registered outside the shellmidden and, especially, the knapping processes are registered in the zones most distant from it (Clemente *et. al.* 2008). The functional analysis has allowed us to identify various productive activities carried out with lithic tools. It is worth mentioning that we have not identified butchering activities or hide processing with lithic tools. Perhaps they used other raw materials (shells, hard woods, bamboo, etc.) for these activities. The plant recourses, with 46% of the lithic tools used, were the most worked, compared to mineral resources (rocks and ceramics) with 29% and animal resources, with only 4%. For the rest (21%) of the tools it was not possible to determine the worked material for sure. As far as the exploitation of plant resources is concerned, wood and other woody plants accounted for 50% of those tools, while 39% are *manos* and fragments of *metate* used for grinding (especially of pumpkins and other edible and/or aromatic plants); the remaining 11% were used for working on other plants, mainly non-woody ones. Of the tools used for working with mineral materials, those that were used in the manufacture of ceramics: spatulas, polishers, smoothers and fragments or chips for scraping and making incisions in the fresh and semi-fresh ceramic paste, are worth special mention. This identified series of tools would indicate that at least a part of the ceramics used in habitation unit number 4 at Karoline was manufactured at the site. The perforators used to perforate rocks, shells and/or bones represent a high percentage of the tools. For this they used small sized fragments which had an abrupt trihedral angle and these were used to produce bead necklaces from those materials that are also present in the archaeological registration of the site (Clemente *et. al.* 2008).

CONCLUSION

The sedimentary of the shellmiddens with its high content of carbonates allows many of the perishable materials in other types of sediments to be conserved in them. Therefore, the possibility of recuperating in shellmiddens bony remains of the animals consumed, as well as vegetable remains such as seeds etc., means that we can consider sites of this type as a basic source for the study of the prehistoric societies who occupied these areas of the Humid Tropics, where the conservation of perishable remains is not very usual. From here its is possible to derive many and important data related not only to the production of food but also to other analyses related to the technological level reached by those societies. However, as we have tried to show throughout this work, this does not imply excavating only the shellmidden part of an archaeological site because it is there where we are going to recuperate most remains.

In the case that we have presented, it is patently clear that many of the productive activities were developed in the southern zone of the excavation, outside the shellmidden. It is there, where most of the work processes related to he production of tools, manufacture of ceramics, woodwork and other plants, took place. However, basic productive activities for subsistence, such as the preparation and/or conservation of food were also carried out on the surface of the shellmidden, since the sediment of the shellmidden was ideal for a quick filtering of the rainwater and allowed the hearths and the different structures related to them to be installed there.

The term "*shellmidden*" (*conchero, shell mounds, amas à coquillage, kjökkenmöddings, sambaqueiras,* etc.) has been used to describe different archaeological phenomena which represent distinct realities and whose only thing in common the presence of shells or valves de molluscs as evidence of the productive processes carried out by a specific social group. This anthropic sedimentation is not identical in all the sites, the variability of the shellmiddens in both shape and size, or because of its composition, is very marked. Moreover, in some cases the shells of the molluscs that were eaten have been used as either constructive elements – as is the case of the ring-shaped forms of the Yaghan huts of Tierra del Fuego, or else to situate structures of hearths, being the most permeable place of the site – as for example Karoline shellmiddens on the Atlantic coast of Nicaragua, or else to contain other structures with other specific functions such as burials – like those found in other sites of Central America and the Caribbean (for example, in Cerro Mangote in Panama or Banwari Tracé in the Dominican Republic (McGimsey 1956, Veloz 1991)

Our short experience in the excavation of these Caribbean shellmiddens allowed us to recognise their sedimentary complexity and has taught us the difficulty in interrelating the part of the shellmidden with the level of occupation outside it, where the stratification is less evident because of the characteristics of the sedimentary matrix and the intensity of some taphonomic processes. However, we considered the necessity for an extensive excavation of these sites as something basic of archaeology itself, which as a social science that seeks to interpret the social relations of the various prehistoric groups from the archaeological register that we created. It is clear that social activities were carried out both inside and outside the shellmidden in habitation unit number 4 at Karoline, and accordingly both zones should be considered worthy study. We can consider the surfaces of these shellmiddens as a dynamic and continuous sedimentation, since hearths were excavated, structures were raised and modified, materials were moved, etc in them. For this reason there are cases where we can find some sediments, for example from the cleaning of the oldest fireplaces, which superimpose more modern sedimentations. That is why it is necessary to be cautious with the materials that we date in these contexts and to know exactly where they come from. The excavations made in shellmiddens by artificial layers and in reduced spaces make it difficult to ensure a level of control that allows us detect and evaluate that phenomenon to which we refer.

Acknowledgements

This work is part of the Proyecto de Investigación (HUM-2005-02268) "*Economía y medio ambiente en la costa Atlántica de Nicaragua: diversidad, complementareidad y cambio en la prehistoria*", financed by the Ministry of Education and Science. The field work has been financed by the Instituto de Patrimonio Histórico (Ministry of Culture) and the Agencia Española de Cooperación Internacional throughout different meetings.

References

CLEMENTE, I., GASSIOT, E. & TERRADAS, X. (2008). Manufacture and use of stone tools in the Caribbean Coast of Nicaragua. The analysis of the last phase of the shellmidden KH-4 at Karoline (250-350 calAD). In: Laura Longo and Natalia Skakun (eds.), *"Prehistoric Technology" 40 years later: Functional studies and the Russian Legacy*. BAR International Series 1783, pp.285-294, BAR Publishing Oxford.

ESTÉVEZ, J & VILA, A (2000). Estratigrafía en contexto, *KREI* 5, 29-61, Vitoria.

GASSIOT BALLBÈ, E. (2005). Shell middens on the Caribbean coast of Nicaragua: Prehistoric patterns of mollusc collection and consumption. Daniella E. Bar-Yosef Mayer (ed.), *Archaeomalacology: Mollusc in former environments of human behaviour*. 9th ICAZ Conference, Durham 2002, pp. 40-53. Oxbow Books, Oxford.

GASSIOT, E. & PALOMAR, B. (2006). Prehistoric settlements of the Atlantic COSAT of Nicaragua. Absolute chronology of Pearl Lagoon and Bluefields shellmiddens. In H.D. Yacobaccio & D.E. Olivera (cords.) *Americam prehistory*. BAR International Series 1524, pp. 233-241, BAR Publishing, Oxford.

GASSIOT, E., CLEMENTE, I. & PALOMAR, B. (2003). Entre lagunas y manglares: poblamiento y explotación del litoral en la Costa Caribe de Nicaragua (1.400 calAC a 1.000 calDC). Paper presented in the International Association of Caribbean Archaeology Congress, 29 de junio-6 de julio. Santo Domingo.

MAGNUS, R. (1974). T*he Prehistory of Miskito Coast of Nicaragua: A Study in Cultural Relationships*. Ph.D. Thesis, Yale University. Ann Arbor: University Microfilms.

MAGNUS, R. (1975). La secuencia cerámica de la Costa Atlántica y la zona central de Nicaragua. *Boletín Nicaragüense de Bibliografía y Documentación* 4, pp. 1-10.

MAGNUS, R. (1976).La Costa Atlántica de Nicaragua. *Vínculos* 2(1), pp. 67-74.

MAGNUS, R. (1978). "The Prehistoric and Modern Subsistence Patterns of the Atlantic Coast of Nicaragua: A Comparison", in Stark, B. and Voorhies, B., eds., *Prehistoric Coastal Apatations. The Economy and Ecology of Maritime Middle America*. New York, Academic Press, 61-80.

MATILLÓ, J. (1993) Los 'conchales' de Punta Mico, in Arellano, J.E., ed., *Treinta años de arqueología en Nicaragua*. Managua, Museo Nacional de Nicaragua, pp. 17-18.

McGIMSEY, CH.R.III (1956). Cerro Mangote: A Preceramic Site in Panama. American Antiquity, vol. 22(2), pp. 151-161.

NIETSCHMAN, B. (1973). *Between Land and Water. The Subsistence Ecology of the Miskito Indians, Eastern Nicaragua*. New York, Seminar Pres.

ORQUERA, L.A. (1996). Túnel VII: la estratigrafía, in J Estévez and A Vila (coords), *Encuentros en los conchales fueguinos*. Treballs d'Etnoarqueologia 1, pp. 83-103, CSIC y UAB, Barcelona.

ORQUERA, L.A. & PIANA, E.L. (1992). Un paso hacia la resolución del palimpesto. En: *Análisis Espacial en la Arqueología*. Compilación L.A. Borrero y J.L. Lanata; ed. Búsqueda de AYLLU S.R.L., pp. 21-52, Buenos Aires.

ORQUERA, L.A. & PIANA, E.L. (1996). Túnel VII: la excavación. En: J. Estévez y A. Vila (coords.), *Encuentros en los conchales fueguinos*. Treballs d'Etnoarqueologia 1, pp. 47-81 CSIC y UAB., Barcelona.

ROACH, J. (1991[1798]). Las sorprendentes aventuras de John Roach, marinero de Whitehaven. *Wani* 11, pp. 14-26.

ROMERO, G. (1995). *Las sociedades del atlántico de Nicaragua en los siglos XVII y XVIII*. Managua, Banco Nicaragüense.

SQUIER, E.G. (1891). *Adventures on the Mosquito Shore*. New York: Worthington Co.

VELOZ, M. (1991). *Panorama histórico del caribe Precolombino*. Banco Central de la República Dominicana. Santo Domingo.

GEOARCHAEOLOGICAL INVESTIGATIONS AT SHELL MOUNDS, SOUTHERN BRAZIL

Marisa Coutinho AFONSO
Museu de Arqueologia e Etnologia/Universidade de São Paulo. São Paulo, Brasil. marisa.afonso@usp.br

Laércio Loiola BROCHIER
Museu de Arqueologia e Etnologia/Universidade de São Paulo. São Paulo, Brasil. PhD student, CNPq

***Abstract**: Brazilian South coast presents a concentration of shell mounds (sambaquis), including the highest one (Garopaba do Sul), 22 meters high. Although these sites have been studied since the nineteenth century, the social characteristics of its formation are still poorly known. Till the 1980's, Brazilian shellmounds were considered products of low demographic populations, bands with high mobility in a territory involving a variety of marine environments, with subsistence based mostly on shellfish. In the 1990's new research on shell mounds, involving zooarchaeological, geoarchaeological, bioanthropological and chronological studies, carried out in the north and south parts of Santa Catarina State coast indicated long-duration occupations by people specialized in exploitation of sea resources (fish and shellfish), using shells to build the sites. Geoarchaeological studies aim to identify sites distribution in the landscape, based on recent geological data on sea-level changes and also to identify site formation processes, based on stratigraphical and sediment analysis and postdeposicional processes. Detailed researches were carried out at two different types of locations (Guaratuba Bay and Babitonga Bay) with large shell mounds visible as landmarks, small ones, and other types of coastal sites, displaying a very complex occupation pattern. This paper presents some sites in different areas as examples of the field methodology applied to shell mounds and data interpretation.*
***Key words**: shell mounds, site formation processes, geoarchaeology, Brazil*

***Résumé**: La côte sud du Brésil présente une concentration des amas coquillers (sambaquis), y inclus le plus haut (Garopaba do Sul), avec 22 metres. Même avec les études de ces types de sites dès le XIXème* siècle, *les caracteristiques sociales de ses formations sont encore pauvrement connues. Jusqu'aux années 1980, les amas coquillers Brésiliens ont été considerés comme des produits de populations de faible densité démographique, bandes de grande mobilité dans un territoire qui présente plusiers environnements, avec une économie de* subsistence basée *surtout sur les coquillages. Dans les années 1990, de nouveaux recherches sur les amas coquillers, qui assemble les études zooarchéologiques, géoarchéologiques, bioanthropologiques et chronologiques, dans les côtes au nort et au sud de l'État de Santa Catarina, montrent des occupations de longue terme des groupes humaines spécialisés dans l'exploration de resources marines (poissons et coquillages), et qui ont utilisés des coquillages pour construire les sites. Les études géoarchéologiques ont l'objectif de identifier la distribuition des sites dans le paysage, fondés sur les plus recentes données géologiques des variations du niveau de la mer et aussi identifier les processus de formation des sites, associer les analyses stratigraphiques, sédimentologiques et les processus post-dépositionnaux. Recherches détaillés ont été ménés sur deux lieux différents (les baies de Guaratuba et Babitonga, Sud du Brésil), où il ya a grands amas coquillers bien visibles, petits amas coquillers et autres types de sites, impliquant ume mode d'établissement complexe. Ce papier présente quelques sites comme examples de l'approche méthodologique développée sur le terrain aux amas coquillers et l'interpretation des données.*
***Mots clés**: amas coquillers, processus de formation de sites, géoarchéologie, Brésil*

INTRODUCTION

Brazilian coast presents a concentration of shell mounds (called "sambaquis"), including the hightest one (22 meters), Garopaba do Sul, at the southern coast of Santa Catarina state. Although shell mounds have been investigated since the nineteenth century, their formation processes are not yet completely understood.

Shell mounds are located along Brazilian coastline from the north of the country (Pará and Maranhão states), where pottery has been found in these sites, with a gap between north and northeast, where shell mounds have not been located, and from the north of the of Bahia state to the south of Rio Grande do Sul state.

Till the 1980's, shell mounds were considered as places occupied by gathering and fishing groups, socially organised as bands, with high territorial mobility, low populational density and subsistence based on collecting mollusks as *Anomalocardia brasiliana*, oysters and mussels. Shell mounds were interpreted as areas for discarding food remains, specially shells from mollusks.

Brazilian shell mounds were built with shells and fish bones; burials, hearths, habitation structures, artifacts (lithic, bone) and considered as places for habitation and for burials (Gaspar 1998).

In the 1990's, new researches were carried out in shell mounds, in the coastal areas of Rio de Janeiro, São Paulo and Santa Catarina states (Southeastern and Southern Brazil) leading to different interpretations about the societies who built these sites.

One of the first subjects that had quite a different interpretation was subsistence. Although when, looking at shell mounds, at stratigraphical profiles, shells are much more visible than fishes' remains, in most part of the shell mounds, zooarchaeological analysis demonstrated that, in fact, fishes represent the most important part of food resources (Figuti, 1992; Bandeira, 1992; Klökler, 2001).

New research carried out in the north and south parts of Santa Catarina state coast indicated long-duration occupations by people specialized in exploitation of sea resources (fish and shellfish), using shells to build the sites.

Studying the site formation processes, shells represented not only the remains of faunal resources but also acted as constructive material for building the shell mounds. Some of them can be considered as landmarks indicating a much more complex social structure and a less important territorial mobility.

Geoarchaeological researches in shell mounds aimed to identify the distribution of sites in the landscape, based on recent geological data about sea level changes and on regional coastal geomorphology, and to identify site formation processes through stratigraphical and sedimentological analysis and also postdeposicional processes.

Detailed researches were carried out at different types of locations: large shell mounds visible as landmarks, small ones, and other types of coastal sites, displaying a very complex occupation pattern, difficult to be recognised by archaeologists, specially the definition of site functions. This paper presents some sites in different areas as examples of the field methodology applied to shell mounds and data interpretation.

GUARATUBA BAY – BABITONGA BAY (SOUTHERN BRAZIL)

Guaratuba Bay is located at the coastal plain of Paraná state (Southern Brazil) and involves an articulated hydrographic system with three important rivers (Cubatão, São João and Boguaçu), besides other rivers from the coastal plain and from the slopes of the mountain area named Serra do Mar (Figure 17.1). The bay has only one narrow connection with the Atlantic Ocean, 500m wide, limited by outcrops.

The shell mounds' spatial and chronological distribution, variation in malacological contents, availability and proximity of mollusks' areas and the geological substratum have been used as indicators of the relative sea level oscillations during the Holocene in Brazil (Bigarella, 1954; Martin *et al.* 1988; Gaspar, 2000). However, more recently, geological researches (Angulo, Lessa & Souza, 2005) showed some difficulties in the interpretation of the direct use of informations acquired from shell mounds. Despite a still little participation of archaeologists in this discussion, these questions stimulate the return of rechearches about the cultural reasons for the settlement, construction processes and postdepositional transformations. Also they allow more research on the understanting of migration processes, mobility and settlement of the groups, aspects of the dynamic and evolution of the coastline, allowing the production of more reliable data to link archaeological and geological studies of the coastal region.

Despite their importance, researches on paleogeographical evolution are still very few (Bigarella, 1946, 1954; Martin *et al.* 1988; Lessa *et al.* 2000 and Souza *et al.* 2001) and most investigations associating coastal and shell mound studies, with a real geoarchaeological approach, were conducted along the southeastern-southern coast.

One of the first references was Krone's study and the association between malacological content in shell mounds and sea level variations. Bigarella since the 1940's used geological and paleogeographical approaches in shell mounds and compared them with coastal evolution in Southern Brazil (Bigarella 1946, 1954). Informations about climatic and sea level oscillations allowed interpretations about chronological aspects, occupation and subsistence of the groups who built the shell mounds and hypothesis about the variability in the composition and in cultural remains found in these sites.

Emperaire and Laming (1954/56) established relative ages for some shell mounds in Paraná and São Paulo states, based on the height of geological substratum in relation to sea level.

Bigarella (1954) associated the location of shell mounds to 5 different kinds of geological substratum: a) sandy areas, b) sandy or sandy-clay islands in the middle of mangroves; c) on rocks; d) in the middle of mangroves and e) in the middle of sandy areas. Martin *et al.* (1988) agreed with Bigarella's perspective and considered that the identification of the substratum could help estimate when the site was constructed.

Hurt and Blasi (1960) presented a complementary classification, based on actual characteristics of sea water dynamics: shell mounds located along the banks of rivers that are influenced by tides, along the bays' water lines; located in islands, and along oceanic beaches. They also analysed aspects about the correlation between altimetric levels of ancient beaches and the construction of shell mounds, believing that changes in shell content were not associated to antiquity, as Krone thought, but were important indicators of ecological zones.

At Paraná State, archaeological surveys carried out by Bigarella, Rauth, Posse, Blasi, Martin, Parellada, Angulo, Chmyz and Brochier recorded 298 sites, with only 13 of them already excavated.

At Guaratuba Bay (south of Paraná state), there are 100 shell mounds, and only 3 had been excavated: Ilha dos Ratos, Araújo II e Matinhos II (Fig. 17.2 and 17.3). Radiocarbon datings indicate that the construction of shell mounds in this area involves a period from 6.500 years B.P. (Sambaqui Ramal, at Paranaguá Bay) to 1.500 anos B.P. (Sambaqui Ilha dos Ratos, at Guaratuba Bay).

Fig. 17.1. Location map of Paraná and Santa Catarina state coastal area (Southern Brazil)

Fig. 17.2. Shell mounds at Guaratuba Bay

Fig. 17.3. Sambaqui Ilha dos Ratos (shell mound located at Guaratuba Bay; Joseph Emperaire and Annette Laming-Emperaire carried out research in the 1950's)

During archaeological surveys, other sites quite different from shell mounds were located like lithic sites, ceramic sites and historical sites (Chmyz, 1975, 1976, 1986).

Some studies about the origin, mobility and dispersal of the groups who built shell mounds were carried out along the Brazilian coast.

Based on radiocarbon datings, Gaspar (1996; 2000) constructed a model of occupation of fisher-hunter-gatherer groups who inhabited the southeastern-southern coast and identified a possible center of dispersion in the coast of Paraná, 6.500 years B.P. The displacement of these groups to other areas would have been very quick, reaching the coast of Rio de Janeiro, to the north, around 5.500 years B.P.

Oliveira (2000) proposed a model about spatiality and routes of dispersal based on the geological study of possible physical conexions between old inland plains.

Fig. 17.4. Shell mound 20 m high located near São João River at the route named "Eixo São João-Palmital" by Oliveira (2000). Testing Oliveira's predictive model based on paleogeographic studies, Brochier located some shell mounds, following this route, a spatial connection between two bays with many shell mounds

Oliveira indicated that the region between São João River (Paraná state) and Palmital Channel (Santa Catarina state, to the south) offered fisiographic controls in the middle Holocene favorable to the displacement of populations between the bays of Guaratuba and Babitonga. He pointed out some natural corridors (like the hydrographic basins of Itapocu and Piraí) that were important to spatial integration between groups from the coast and groups from the higher and inland areas (Terras Altas).

Oliveira (2000) located 42 shell mounds at Babitonga Bay including the largest one (sambaqui Tiburtius), 18 m high and 92.000 m^3. He also identified the geological substratum of all the shell mounds located at the coastal plain of Joinville: 14% are located on bedrock, 12% on fluvial-lagoonal deposits, 34% on alluvial fan deposits and 40% on eolian deposits.

Analysing the altitude of the shell mounds' base and the relative sea level curves, Oliveira (2000) observed that the first occupation of the major part of the sites in the area happened between 5.100 and 3.600 year BP, when mangroves were expanding. He also proposed a route named "Eixo São João/Palmital" that helped the displacement of the population between two bays (Guaratuba and Babitonga), where Brochier located many shell mounds including one 20 m high (Fig. 17.4 and 17.5).

Bentz (2001) organised a statistical analysis of the dimensions of these 42 shell mounds and noticed that they are usually between 1 and 10 m high, with length and width between 20 and 100 m. Sites higher than 14 m and with length and width more than 180 and 140 m are less frequent. The dimensions of shell mounds can be associated to natural elevations.

Another explanatory proposition based on geoarchaeological research in a regional scale is being tested by Brochier at Guaratuba Bay, directed to the mapping of "coastal geoarchaeological controls" (Brochier, 2001). These controls indicate natural restrictions based on relationships between geomorphic processes able to modify or control preservation, crono and morpho stratigraphical distribution, visibility and accessibility, erosion or deposition of sites. If not considered, these geoarchaeological restrictions may lead to misinterpretations; if considered, they may guide methodological aspects and point out favorable situations to the recovery of preserved remains, verify the effect of erosive or deposicional processes in the site settlement pattern and indicate geological, paleogeographical and paleoclimatic contexts capable of guiding cultural choices, like the mapping of possible geoecological corridors of displacement, occupation and mobility (as the route Palmital-São João model).

Brochier (2001) indicated an example of the use of geoarchaeological controls: the possibility of finding old sites, from Middle and Lower Holocene, of coastal adaptation in geomorphic and sedimentary traps more than 10 meters above the surface, still on the actual coastal plain. These traps would be located near dissecated old valleys, formed during the regressive period, and that were transformed in estuaries or rivers with tidal influence (during the next transgressive period), in large extension on the land.

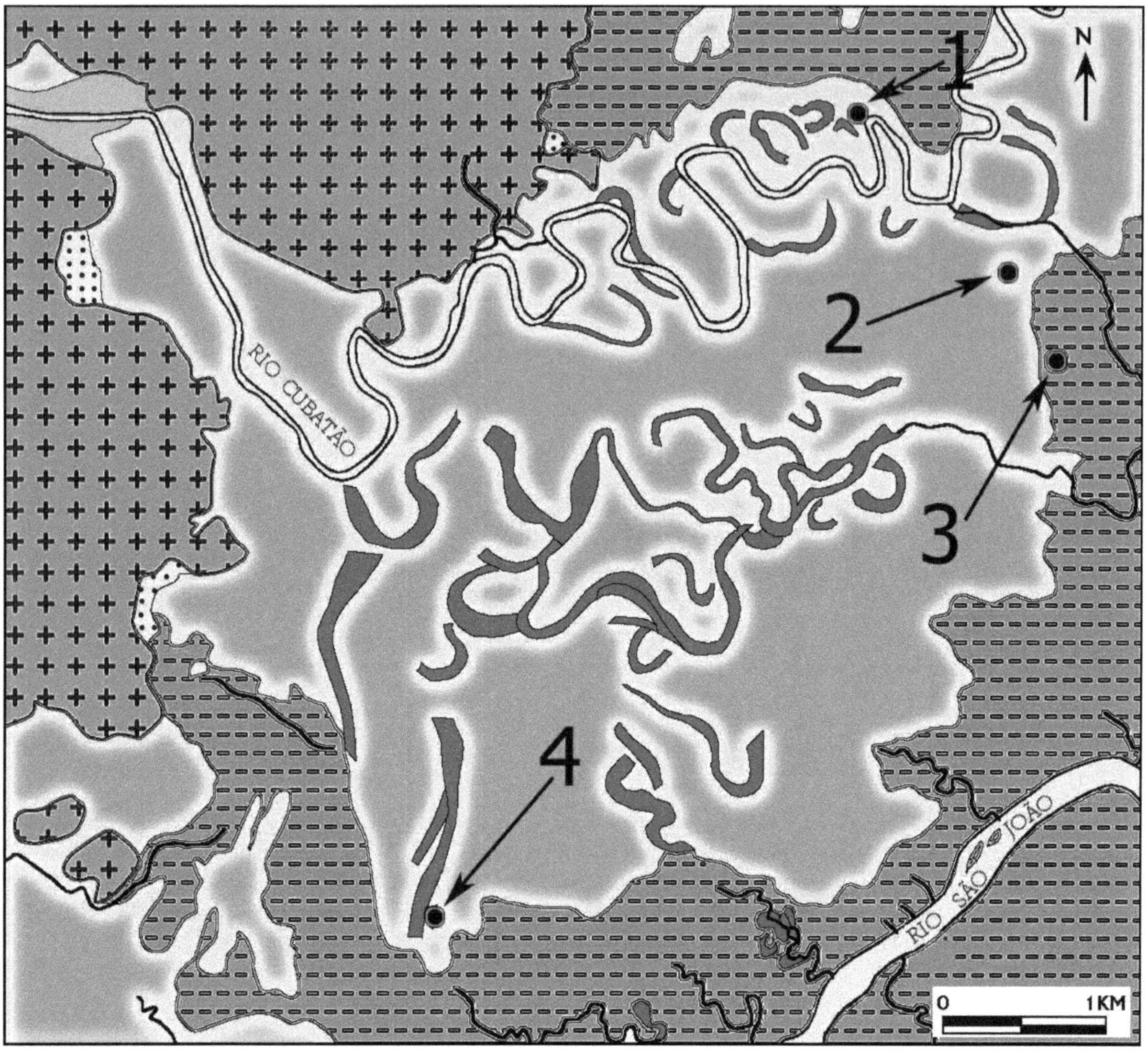

Fig. 17.5. Alluvial fan deposits of Cubatão River (Guaratuba Bay) with the location of shell mounds. 1) 12 m high, partially destroyed by an old meander from Cubatão River, now an oxbow lake; 2) shell mound on colluvial-alluvional deposits, partially buried; 3) 4 m high, on paleoestuarine deposits, well preserved; 4) 20 m high on the alluvial fan deposits or bedrock, near one of the abandonned channels of Cubatão River. These 4 shell mounds are associated to different geomorphic and stratigraphic controls and can be connected to the evolution of the alluvial fan. Geologic map (Angulo 1992)

Coastal sites formed in adjacent areas to these submerged valleys could be covered by estuarine muds or colluvial-alluvial deposits, in specific topographic, climatic and depositional situations that allowed their preservation. This proposition is based on paleogeographical and stratigraphical models to the coastal areas of Paraná and Santa Catarina states (Lessa & Angulo, 1998; Souza *et al*, 2001), on dates from coastal sedimentary deposits and holocenic alluvial deposits and also on paleoecological studies, that indicate sea entrance in valleys during the late Holocene. Although some researchers believe that evidences from old coastal sites, like shell mounds, were destroyed due to the continuous submersion of the continental platform in postglacial periods, Brochier (2001) considers that it will be possible to locate inland sites of 8.000 years or older. This prediction could be possible associating informations from the core profiles from the contact between Holocene estuary muds and Pleistocene barriers or continental substratum, in association to relative sea level curves for the limit Holocene-Pleistocene.

In Babitonga Bay (Joinville and São Francisco Island), some projects were carried out to understand the huge shell mounds formation processes (De Blasis & Afonso 2000). At Sambaqui Espinheiros II, Afonso & De Blasis (1994) identified a thick layer (1 meter) of *Anomalocardia brasiliana*, charcoal, fish bones and preserved plant fibers on the substratum composed by alluvial fan deposits, representing the beginning of shell mound construction and the settlement of population, 2970 ± 60 years B.P. (Fig. 17.6 and 17.7).

Fig. 17.6. Sambaqui Espinheiros II (Babitonga Bay)

Fig. 17.7. Sambaqui Espinheiros II – stratigraphic profile

Stratigraphical and zooarchaeological investigations (carried out by Figuti & Klökler 1996) showed the predominance of only one activity, gathering *Anomalocardia brasiliana*, at the basal and oldest levels, and at the more recent layers products of other activities with fishing as the main one. Gathering of mollusks was a complementary activity: mussels (*Mitella sp*) and oysters (*Ostrea sp., Crassostrea rhizophorea*) from mangroves and *Anomalocardia brasiliana* from mud areas.

Two hearths with very burned and fragmented mussel shells, charcoal and ashes where dated (1160 ± 45 years B.P. and 1270 ± 60 years B.P.) representing the most recent occupation of shell mound builders in the area. At the time of the first European arrivals in Brazil, there were no more shell mound builders.

The almost predominance of one type of mollusk *(Anomalocardia brasiliana)* and the preservation of braided plant fibers indicate a rapid accumulation of shells and that these remains were used as construction material for building a platform, a dry platform on the alluvial fan deposits. Faunal remains were not only related to subsistence but also to build the site.

A geoarchaeological investigation of the sediments was carried out by Bentz (2001). She collected sedimentological samples along a stratigraphic profile at Lagoa do Saguaçu shell mound, near Babitonga Bay. Sediments were composed basically of clay and silt, despite the difference in colour and amount of shells noticed on the profile. Sand incorporates also large amounts of fragmented shells probably originated during the construction of the shell mound. So, despite the differences in texture, color, quantity and fragmentation of the shells, sedimentary matrix presented a similar proportion of fine and coarse sediments.

Geoarchaeological and zooarchaeological investigations in field present different results from the laboratory analysis. Another problem is to locate the profiles and sample collections in order to understand the variability in sediments of a same shell mound and that they represent events and processes that happened in the site. The stratigraphical complexity of shell mounds is very high, posing methodological challenges to choose sample areas and to analysis of the sediments.

Geoarchaeological researches in the bays of Guaratuba and Babitonga involve aspects related to paleogeography, to distribution of sites in the landscape and shell mound formation processes.

Acknowledgments

We thank the following institutions: Museu de Arqueologia e Etnologia/Universidade de São Paulo (MAE/USP) and Museu Arqueológico de Sambaqui de Joinville (MASJ). The research was supported by FAPESP, CNPq and CAPES.

References

AFONSO, M.C. (1999) O sambaqui Espinheiros II: uma contribuição à arqueologia do litoral norte de Santa Catarina. Revista do CEPA, vol. 23, n. 29, pp. 118-123.

AFONSO, M.C.; DE BLASIS, P. (1994) Aspectos da formação de um grande sambaqui: alguns indicadores em Espinheiros II, Joinville, Revista do Museu de Arqueologia e Etnologia, 4, pp. 21-30.

ANGULO, R.J. (1992) Geologia da planície costeira do Estado do Paraná, Tese de doutoramento, Instituto de Geociências/Universidade de São Paulo. São Paulo.

ANGULO, R.J.; LESSA, G.C.; SOUZA, M.C. (2005) A critical review of the mid- to late Holocene sea-level fluctuations on the eastern brazilian coastline. Quaternary Science Reviews, Londres, p. 1-21.

BANDEIRA, D.R. (1992) Mudança de estratégia de subsistência. O sítio arqueológico Enseada I – um estudo de caso. Dissertação de mestrado. Universidade Federal de Santa Catarina.

BENTZ, D. (2001) Os sambaquis da região de Joinville (SC) no contexto da variação do nível relativo do mar. Trabalho Final, Instituto de Geociências/Universidade de São Paulo.

BIGARELLA, J.J. (1946) Contribuição ao estudo da planície litorânea do Estado do Paraná. Arquivos de Biologia e Tecnologia, 1:75-112. Curitiba.

BIGARELLA, J.J. (1950/51) Contribuição ao estudo dos sambaquis no Estado do Paraná II, regiões adjacentes à baía de Guaratuba. Arquivos de Biologia e Tecnologia, 5/6:293-314. Curitiba.

BIGARELLA, J.J. (1954) Os sambaquis na evolução da paisagem litorânea sul-brasileira. Arquivos de Biologia e Tecnologia, 9:199-221. Curitiba.

BROCHIER, L.L. (2001) O uso de controles geoarqueológicos: perspectivas para o estudo e caracterização de áreas arqueológicas na Baía de Guaratuba-PR. Congresso da Sociedade de Arqueologia Brasileira, Rio de Janeiro.

BROCHIER, L.L. (2004) Diagnóstico e manejo de recursos arqueológicos em Unidades de Conservação: uma proposta para o litoral paranaense. Dissertação (Mestrado) – Faculdade de Filosofia, Letras e Ciências Humanas, Universidade de São Paulo, São Paulo.

CHMYZ, I. (1975) A ocorrência de sítio arqueológico com pontas-de-projétil no litoral paranaense. Nota prévia sobre o sítio PR P 31: Ribeirão. Anais da Academia Brasileira de Ciências, Rio de Janeiro – RJ, v. 47, p. 81-89.

CHMYZ, I. (1976) A ocupação do litoral dos Estados do Paraná e Santa Catarina por povos ceramistas. Estudos Brasileiros, 1(1):79-143. Curitiba.

CHMYZ, I. (1986) A formação de sambaquis em período histórico no Estado do Paraná. Arqueologia, Revista do Centro de Estudos e Pesquisas Arqueológicas, Curitiba – PR, v. 5, p. 103-111.

DE BLASIS, P.A.D.; AFONSO, M.C. (2000) Indicadores de complexidade nos grandes sambaquis do litoral sul do Brasil: o caso de Espinheiros II, Joinville. In: Alicia Durán Coirolo e Roberto Bracco Boksar (ed.) Arqueologia de las Tierras Bajas. Montevideo, p. 343-352.

EMPERAIRE, J.; LAMING, A. (1954/1955) Les sambaquis de la côte méridionale du Brésil, campagnes de fouilles. Journal Société des Americanistes, 45:5-163. NS, Paris.

FIGUTI, L. (1992) Les sambaquis COSIPA (4200 à 1200 ans BP):étude de la subsistance chez les peuples préhistoriques de pêcheurs-ramasseurs de bivalves de la côte centrale de l'état de São Paulo, Brésil. Thèse de Doctorat, Muséum National d'Histoire Naturelle, Paris.

FIGUTI, L.; KLOKLER, D.M. (1996) Resultados preliminares dos vestígios zooarqueológicos do sambaqui Espinheiros II (Joinville, SC). Revista do Museu de Arqueologia e Etnologia, 6: 169-187.

GASPAR, M.D. (1998) Considerations of the *sambaquis* of the Brazilian coast. Antiquity 72: 592-615.

GASPAR, M. (2000) Sambaqui: Arqueologia do litoral brasileiro. Editora Zahar.

HURT, W.; BLASI, O. (1960) O Sambaqui do Macedo: A.52.B. Paraná, Brasil. Arqueologia, Curitiba, (2):1-98.

KLOKLER, D.M. (2001) Construindo ou deixando um sambaqui? Análise de sedimentos de um sambaqui do litoral meridional brasileiro: processos formativos, região de Laguna, SC. Dissertação de Mestrado. Faculdade de Filosofia, Letras e Ciências Humanas/ Universidade de São Paulo.

LESSA, G.C.; ANGULO, R.J. (1998) Oscillations Or Not Oscillations, That Is The Question – Reply. Marine Geology, Amsterdan, v. 150, p. 189-196.

MARTIN, L.; SUGUIO, K.; FLEXOR, J.M.E AZEVEDO, A.E.G. (1988) Mapa geológico do quaternário costeiro dos estados do Paraná e Santa Catarina com texto explicativo, Boletim do DNPM, n.18, 40p. (série geologia básica). Brasília.

OLIVEIRA, M.S.C. (2000) Os sambaquis da planície costeira de Joinville, litoral norte de Santa Catarina: Geologia, Paleogeografia e Conservação in situ. Dissertação de Mestrado, Universidade Federal de Santa Catarina.

SOUZA, M.C.; ANGULO, R.J.; PESSENDA, L.C.R. (2001) Evolução paleogeográfica da planície costeira de Itapoá, litoral norte de Santa Catarina. Revista Brasileira de Geociências, São Paulo, v. 31, n. 2, p. 223-230.

Session WS32

INTERDISCIPLINARY STUDIES IN HUMAN EVOLUTION

INTRODUCTION

Edited by:
Grupo de Estudos em Evolução Humana
Departamento de Antropologia, Universidade de Coimbra, Rua Arco da Traição, 3000-056 Coimbra.;
E-mail: greevh@gmail.com; Web site: www.geevh.org

Eugénia CUNHA

INTRODUCTION

> *"A palavra ciência, no sentido em que a usamos hoje, é uma invenção do séc. XIX. Mas o pensamento científico existe desde que existe a invenção da linguagem, e, com ela, a formulação das primeiras perguntas do homem sobre os mistérios do universo enorme e sem nome que o rodeia. Estas perguntas são tão poderosas na dinamização da estruturação de ideias como as primeiras organizações de mecanismos elementares de defesa contra um mundo imediatamente circundante e perigoso, onde só a inteligência pode permitir a sobrevivência de um primata glabro, sem garras e sem dentes de sabre, ou qualquer outro equipamento natural que jogue directamente a seu favor…"*
>
> (Avelar; Matos; Rego, 2004:7)

The subject of Human Evolution is complex, obligatorily interdisciplinary, constituting a fascinate challenge that requests permanent updates.

All science lives in a constant mutation. However, and in what concerns the "satellite" disciplines that rotate around this "planet" that is the Human Evolution, the progresses are extraordinary fast, and frequently, the discovery of a small fossil may signify a return of millions of years in the dates attributed to the first hominines.

This avalanche of knowledge, conjugated with the complexity of some fields, such as Genetics (today an essential tool to understand our own historic evolution) and Molecular Biology, has produced a truly interdisciplinary movement that claim to the interception of several scientific domains. Taking into account this perspective, and recognizing the importance of interdisciplinary as a way to achieve knowledge in a less narrow perspective about the reality that surround us, we have proposed the organization and coordination of a workshop focussed in the disciplines that enrich the cognitive field of Human Evolution.

INTRODUCTION

> *"A palavra ciência, no sentido em que a usamos hoje, é uma invenção do séc. XIX. Mas o pensamento científico existe desde que existe a invenção da linguagem, e, com ela, a formulação das primeiras perguntas do homem sobre os mistérios do universo enorme e sem nome que o rodeia. Estas perguntas são tão poderosas na dinamização da estruturação de ideias como as primeiras organizações de mecanismos elementares de defesa contra um mundo imediatamente circundante e perigoso, onde só a inteligência pode permitir a sobrevivência de um primata glabro, sem garras e sem dentes de sabre, ou qualquer outro equipamento natural que jogue directamente a seu favor…"*
>
> (Avelar; Matos; Rego, 2004:7)

La discipline de l'Evolution Humaine est complexe, interdisciplinaire, et constitue un défie fascinante, demandant des actualisations en permanence.

La science est en perpétuel changement. Cependant, et en ce qui concerne les disciplines autour de l'évolution humaine, les progrès sont extraordinaires, rapides, et souvent la découverte d´un petit fossile mène a des reculs temporels de l'ordre des *millions d'années* dans les datations attribués aux premiers hominiens.

Une telle avalanche de connaissance en articulation avec la complexité des certains thématiques, comme la Génétique (aujourd´hui un instrument essentiel pour la compréhension de notre histoire évolutive), et la Biologie Moléculaire, à déclenché un véritable mouvement interdisciplinaire que demande la communication entre plusieurs demains scientifiques.

En essayant de valoriser l´importance de l´interdisciplinarité comme outil d'acquisition de connaissance, on a proposé la réalisation d'un workshop dirigé vers ces sciences "satellites" de l´ Evolution Humaine.

INTRODUÇÃO

> *"A palavra ciência, no sentido em que a usamos hoje, é uma invenção do séc. XIX. Mas o pensamento científico existe desde que existe a invenção da linguagem, e, com ela, a formulação das primeiras perguntas do homem sobre os mistérios do universo enorme e sem nome que o rodeia. Estas perguntas são tão poderosas na dinamização da estruturação de ideias como as primeiras organizações de mecanismos elementares de defesa contra um mundo imediatamente circundante e perigoso, onde só a inteligência pode permitir a sobrevivência de um primata glabro, sem garras e sem dentes de sabre, ou qualquer outro equipamento natural que jogue directamente a seu favor..."*
>
> (Avelar; Matos; Rego, 2004:7)

A temática da Evolução Humana é complexa, obrigatoriamente interdisciplinar, constituindo um desafio fascinante que requer actualizações constantes.

Toda a ciência vive em mutação constante. No entanto, e no que diz respeito às disciplinas "*satélite*" que giram em torno deste "*planeta*" que é a Evolução Humana, os avanços são extraordinários, velozes e, muitas vezes, a descoberta de um pequeno fóssil têm significado um recuo temporal de *milhões de anos* nas datas atribuídas aos primeiros hominíneos.

Esta avalanche de conhecimentos conjugada com a complexidade de algumas temáticas, como é o caso da Genética (hoje uma ferramenta essencial da compreensão da nossa história evolutiva), da Biologia Molecular tem gerado um verdadeiro movimento interdisciplinar que reclama a comunhão de vários domínios científicos. Tendo por base este pressuposto, e sobrelevando a importância da interdisciplinaridade enquanto veículo de aquisição de um conhecimento mais abrangente e menos redutor da realidade, propusemos a organização e coordenação de um workshop dirigido às ciências que alimentam o campo cognitivo da Evolução Humana.

HAMMERS, ANVILS & NUTS: CHIMPANZEES TECHNOLOGY? APPLYING THE CONCEPT OF "*CHAÎNE OPÉRATOIRE*" TO "*NUT-CRACKING*": INTERDISCIPLINARY RESEARCH

S. CARVALHO

Group of Studies in Human Evolution, Department of Anthropology, University of Coimbra, Portugal

T. MATSUZAWA

Primate Research Institute, Kyoto University. Japan

C. SOUSA

Department of. Anthropology, Faculty of Social and Human Sciences, New University of Lisbon, Portugal

E. CUNHA

Department of Anthropology, University of Coimbra. Portugal

Abstract: *"Nut-cracking" behaviour in natural habitat has been reported since 1951, in different communities of chimpanzees and today it's recognised as cultural behaviour. Recent study indicate that chimpanzees recognize the function of a tool and adapt it to execute more than just its function, actually transporting it, substituting an original raw material and reusing it, if necessary, in order to optimize a task. For this project we combine primatological and archaeological methodologies during fieldwork in Guinea and attempted to detect an operational sequence by means of direct and indirect observation, during the practice of nut-cracking by wild chimpanzees. Do we have a "Chaîne Opératoire" in this process? Can we talk about "nut-cracking" typology and technology? These questions involve a complex cognitive capacity witch is, often, exclusively related with human primates.*

Key-words: *human evolution; interdisciplinary; nut-cracking; lithic material; chaîne opératoire; chimpanzee*

Résumé: *Le comportement du "nut-cracking" -cassage des noix- dans l'habitat naturelle a été reporté depuis 1951, en différentes populations de chimpanzés et aujourd'hui est reconnu comme un comportement culturel. Des études récentes indiquent que les chimpanzés reconnaissent la fonction d'une outil et adapte le pour exécuter plus que sa fonction, en réalité en le transportant, remplacent une matière première initial et le réutilisant, si nécessaire, de façon a optimisé un travail. Pour ce projet on rejoint méthodologies de l'archéologie et de la primatologie pendant les travaux réalisent en Guinée Conakry. On prétend détecter l'existence d'une chaîne opératoire, avec de l'observation direct et indirect, dans la pratique de "nut-cracking" pour chimpanzés, en habitat naturelle. Il y a une "chaîne opératoire" pendant ce procès? Il est possible parler d'un typologie et technologie en rapport au comportement "nut-cracking"? Ces questions remettent pour une capacité cognitive complexe qui est, quelquefois, exclusivement considérée comme propre des primates humains.*

Mots-clés: *Evolution humaine; interdisciplinaire; nut-cracking; matériel lithique; chaîne opératoire; chimpanzés*

INTRODUCTION: THE *CHAINE OPERATOIRE*

Application concepts

"The development of this ability to manipulate branches and twigs has led to the utilisation of sticks as tools, in the Gombe Stream chimpanzee population...In breaking a twig from a tree and then stripping it of its leaves to suit it for a specific purpose, the chimpanzee is not simply utilizing a natural object, but actually making a very simple tool" (Goodall, 1963: 44).

"Man is a social animal, distinguished by culture: by the ability to make tools and communicate ideas" (Oakley, 1965:1).

The *Chaîne Opératoire* theory and method was originally introduced by ethnographical observation, serving as a medium to describe traditional techniques. Mauss introduced technology as a phased process that can be analysed through several repeated stages, beginning with the raw material selection and concluding with the discarding of the tool (Mauss, 1967). Although this theory was already archaeologically developed in 1964-65 (Leroi-Gouhran, 1964), the first applications of this concept to archaeology date back to the 1970's in Leroi-Gouhran's works on the upper Palaeolithic site of Pincevent (Leroi-Gouhran and Brézillon, 1972). This concept, applied to prehistoric archaeology, allows the researcher to identify, on a lithic assemblage, different phases of blank production leading to tool construction and use (Boëda *et al.*, 1990; Lucas, 2000). For each stage, it is important to identify *which* raw material or *which* tool is produced, but equally vital is to know *how* the raw material is acquired or *how* the tool is manufactured. Different strategies can tell us more about variability concerning lithic assemblages and, consequently, can justify cultural variations between related or neighbouring groups of populations (Perlès, 1992).

The analysis of a reduction strategy (operational sequence) implies using an analytical process that encompasses the diverse phases and components of the production and use of an artefact. Not only is the artefact analysed, but also all the actions that contributed to its creation. This allows the researcher to combine the technological study of the artefact, achieved through the reconstitution of its operational sequence, with a more traditional typological study. These two points of view, previously separated from each other, have become unified and complementary.

Ethnographers and archaeologists use, occasionally, the same concepts for analysis, but their research is based on studying different data. The ethnographer can access the whole technical process through direct observation, while the archaeologist is constrained by the remaining objects that represent a frame of the past action, a restrained source of information. Thus, archaeologists should lend methods from other disciplines like ethnography as these enlarge the possibility of organizing the collected data, increasing the range of interpretation hypothesis concerning the technical process (Lucas, 2000).

In archaeology, however, despite the greater accuracy afforded by the study of operational sequences in inferring how a certain community exploits and manages a territory, the methodology implemented cannot fill all the existent gaps. These gaps arise from the passage of time and the irrecoverable loss of essential information.

Ethnographers and archaeologists use, occasionally, the same concepts for analysis, but their research is based on studying different data. The ethnographer can access the whole technical process through direct observation, while the archaeologist is constrained by the remaining objects that represent a frame of the past action, a restrained source of information (Lucas, 2000). Methodological appropriations from other disciplines such as ethnography are fundamental for archaeologists (Lucas, 2000:24) and also for primatologists (Wynn and McGrew, 1989), as they allow one to organize available information and to improve the interpretation of technical processes.

Primatology: a new subject

In Primatology, the study approach appears to be systematically centred on the use of tools, reflected by the constant use of the term "tool-use" in the literature. The tool was initially considered a physical extension of the primate (Goodall, 1964). Other more inclusive definitions, like those of Matsuzawa and Sousa, described and enlarged the various types of "tool-use", thus attributing more complex connotations to them (Matsuzawa, 1999; Sousa, 1999). However, even today, one rarely carries out a study approach that separates the tool from its user, considering it to be an individual entity with its own technological history, reflecting a set of the social and economical constraints of a specific group.

Presently, long-term studies conducted in various areas of Africa reinforce the idea that each community of chimpanzees develops its own set of cultural traditions (Matsuzawa *et al.*, 2001). Many times these traditions reveal themselves very elaborated, not only by means of *tool-using,* but also in the very existence of norms and patterns of use of these tools (Goodall, 1964; Matsuzawa, 1991; Sousa, 1999; Whiten *et al.*, 1999). One observes, in fact, a deliberate modification and use of one or more objects for a specific purpose.

Furthermore, chimpanzees in different communities seem to have diverse cultural traditions, which are passed on from one generation to the other (Goodall, 1964; Nishida, 1987; Boesch, 1996; Whiten *et al.*, 1999; Whiten and Boesch, 2001; Matsuzawa, 1994; Matsuzawa *et al.*, 2001; McGrew, 2002; Biro *et al.*, 2003; McGrew, 2004; Biro, Sousa and Matsuzawa, 2006). Bossou, for example, seems to be exemplary of a *cultural area* due to use of *stone tools* for *nut-cracking*. The type of nut cracked in each community seems to be determined less by natural ecological factors and more by cultural traditions established therein (Matsuzawa *et al.*, 2001; Matsuzawa, 2002; McGrew, 2002; Biro *et al.*, 2003; Humle and Matsuzawa, 2001; McGrew, 2004; Ohashi, 2006; Biro, Carvalho, and Matsuzawa, in press).

The foremost constraint on the archaeologist in trying to understand an object with its function is not being able to witness its use directly. Despite the improvements offered by use-wear and refitting, many hypotheses remain simply inferred or difficult to confirm (McGrew, 2004). Not being able to make direct observations, archaeology must seek clues to develop the understanding of the ways in which lithic tools were used through the analysis of operational sequences within living communities of non-human primates. Primatology can, from the analysis of an operational sequence, look for a systematic and patterned selection of a specific raw material; a landscape that is exploited to obtain the desired resource; whether or not that resource has similar and systematically identifiable characteristics; whether or not the raw material is always used in the same way; whether or not there is a real technological pattern or a functional and motor standardization reflected in tool use behaviour.

INTERDISCIPLINARY ADVANTAGES: PAST, PRESENT AND FUTURE

> *"At around 2.3 million years ago in the fossil record the first stone tools appear, but that is a relatively recent event since our hominid ancestors probably first appeared at least 7 million years ago (and, since chimps use tools made of many materials, it is highly probable that our common human-ape ancestor used them long ago before 7 million years). What were they using? And who was using them?"* (Hart & Sussman, 2005:214).

The present study, combining primatology and archaeology, can be considered an interdisciplinary *tool,* and strives for a new insight on human evolution and an *ethoarchaeology* research: *the study of the behaviour or behavioural products of living nonhuman species as proxies for extinct ones* (McGrew, 2004:37).

There is no primatological study on the use of tools by non-human primates which elaborates on the existence (or lack of it) of a process that can be considered a pattern from the technological view point. This implies a complex cognitive capacity which is most often exclusively associated with human primates. Quoting Vieira (1997:13), "*the occurrence, little over 2,5 millions of years ago in Africa, of semi-standartized lithic industries, and of distinctly standardized ones, with the Oldowan (2 to 1,8 millions of years ago), marks a new stage in the cultural evolution of the hominids*".

Therefore, only with a systematic study that can discern between the various steps of the process and that focuses on the technological aspects of production and the usage of tools, can we reach a conclusion on this hypothesis, for non-human primates; to ultimately know whether or not there is standardization process. The fact that we might think that some behaviour associated with the production/ use of tools by non-human primates might somehow be automatic or opportunistic does not conflict with the central idea of this research. The first hominins might very well have manufactured and used some materials by pure opportunism. It is likely that it started by a combination of factors, by which necessity and opportunism must have played a fundamental role. Among the studies that focus on the possible strategies of "*tool-using*" in the Oldowan technology practiced by hominins, the prevailing model presents this technology as a simple strategy for solving problems, opportunistic in nature and with a low energy investment in selecting raw materials (Bunn *et al.*, 1980; Isaac *et al.*, 1981; Toth, 1982, 1985 *in* Kimura, 1999). More recently, new studies by Kimura (1999) based on several archaeological sites about 1,5 million years old, reveal a flexible and dynamical strategy that adapts to the type of raw materials available. We believe these two models do not contradict each other; opportunism consists of using efficiently what nature has to offer. The low energy investment associated with this simple technology might have been an engine for evolutionary success, as by keeping energy costs low, it solved the existing problem efficiently and thus attained the desired goal, which is what is expected of any efficient technology.

Therefore, we believe the distinction should be set between the beings that can or cannot transform and get the most out of the raw materials that nature provides, using them with a preset goal.

The conclusions of Kimura's study (1999:807), stating that "*the Oldowan tool-using strategy was dynamic and flexible, in response to changes in raw material availability*", are all the more interesting if we take into account other experiments analyzing "*tool-using*" strategies with the "*nut-cracking*" of Bossou (Sakura & Matsuzawa, 1991), which showed evidence of flexibility and dynamics in the strategy used by the chimpanzees to accomplish their goal: cracking the nut for consumption.

This reveals to us that it is fundamental confirm the hypothesis that there is an operational sequence that demands mental anticipation of tasks. Hence, even if for an analysis of "nut-cracking" the distances between collection and selection of the raw material are minimal, and although the remaining process of the operational sequence seems to us very short, it is nevertheless an unchangeable process in the sense that it is made up of several mental and motor stages, susceptible to being observed, separated, as well as interrelated. There can be distinct sequences to accomplish the same preset goal (proving flexibility and variability), of which neither can be inverted nor can any stage be skipped, under penalty of not reaching the desired goal.

An operational sequence will remain a sequence even if we feel inclined to imagining the whole process, from the point when the material is gathered to that when it is discarded, as one that has precise stages, each lasting a perfectly determined period of time, when, in fact, we know such is not always the case. Depending on necessity and opportunity, the operational sequence is relatively quick and made up of stages than can be more clearly defined or less so; yet the purpose of this research remains the same: to investigate whether or not there is standardization in the process. Notice that the analysis/reconstitution of the operational sequences related to the manufacture and use of tools in prehistoric human communities is not always straightforward. Some instruments indicate a longer and more perfected process, using better materials, while others reveal the use of the only existing resources or that the "craftsman" was not as talented for the task; still, the operational sequence exists as a mental scheme in either case and is, from the conceptual viewpoint, independent from the variations imposed by the environment. Hence, the analysis of its existence is important, as is determining how it occurs.

In terms of human evolution, the existence of systematized operational sequences among non-human primates would have important implications towards the understanding of their behavioural complexity. A confirmation of this hypothesis would contribute to the current debate around the existence of certain parallels between the evolution of material cultures and human biological evolution.

References

BIRO, D.; INOUE-NAKAMURA, N.; TONOOKA, R.; YAMAKOSHI, G.; SOUSA, C.; MATSUZAWA, T. (2003) – Cultural innovation and transmission of tool use in wild chimpanzees: evidence from field's experiments. *Animal Cognition*. 6: p. 213-223.

BIRO, D.; CARVALHO, S.; MATSUZAWA, T. Tools, traditions, and technologies: Interdisciplinary approaches to chimpanzee nut-cracking. In Elizabeth V. Lonsdorf, Stephen R. Ross And Tetsuro Matsuzawa, eds. – The *Mind of the Chimpanzee: Ecological and Experimental Perspectives*. Chicago: Chicago University Press, In Press.

BOËDA, E.; GENESTE, J-M.; MEIGNEN, L. (1990) – Identification de chaines operatoires lithiques du paléolithique ancient et moyen. *Paleo*. 2: p. 43-80.

BOESCH, C. (1996) – The emergence of cultures among wild chimpanzees. In Runciman, W., Maynard-Smith, Dunbar, R., eds. – *Evolution of social behaviour patterns in primates and man*. Oxford: Oxford University Press, p. 251-268.

GOODALL, J. (1963) – Feeding behaviour of wild chimpanzees. *Symposia of the Zoological Society of London*. 10: p. 39-48.

GOODALL, J. (1964) – Tool-using and aimed throwing in a community of free-living chimpanzees. *Nature*. 201: p. 1264-1266.

HART, D.; SUSSMAN, R.W. (2005) – *Man the hunted – Primates, predators and human evolution*. Westview Press: Boulder, CO.

HUMLE, T; MATSUZAWA, T. (2001) – Behavioural diversity among the wild chimpanzee populations of Bossou and neighbouring areas, Guinea and Cote d'Ivoire, West Africa: A preliminary report. *Folia Primatologica*. 72: 2, p. 57-68.

KIMURA, Y. (1999) – Tool-using strategies by early hominids at Bed II, Olduvai Gorge, Tanzania. *Journal of Human Evolution*. 37: p. 807-831.

LUCAS, G. (2000) – *Les industries lithiques du Flageolet I (Dordogne): approche économique, technologique, fonctionelle et analyse spatiale*. Thèse de Doctorat présentée a L'Université Bordeaux I. Bordeaux : Université Bordeaux, Volume I.

MATSUZAWA, T. (1991) – Nesting cups and metatools in chimpanzees. *Behavioral and Brain Sciences*. 14: 4, p. 570-571.

MATSUZAWA, T. (1994) – Field experiments on the use of stone tools by chimpanzees in the wild. In Wrangham, R.W.; Mcgrew, W.C.; De Waal, F.B.M.; Heltne, P.G., eds. – *Chimpanzee Cultures*. Cambridge: MA Harvard University Press, p. 351-70.

MATSUZAWA, T. (1999) – Communication and tool use in chimpanzees: Cultural and social contexts. In Hauser, M.; Konishi, M., eds. – *The design of animal communication*. Cambridge: MIT Press, p. 645-671.

MATSUZAWA, T.; BIRO, D.; HUMLE, T.; INOUE-NAKAMURA, N.; TONOOKA, R.; YAMAKOSHI, G. (2001) – Emergence of culture in wild chimpanzees: education by master-apprenticeship. In Matsuzawa, T., eds. – *Primate origins of human cognition and behavior*. Tokio: Springer-Verlag, p. 557-574.

MATSUZAWA, T. (2002) – *Chimpanzees of Bossou and Nimba 1976-2001 – Collection of articles in English*. Kyoto: Kyoto University.

MAUSS, M. (1967) – *Manuel d'Ethnographie*. Coll. Petite bibliothèque. Payot: Paris.

MCGREW, W.C. (2002) – *Chimpanzee material culture – Implications for human evolution*. Cambridge: Cambridge University Press.

MCGREW, W.C. (2004) – *The cultured chimpanzee: Reflections on cultural primatology*. Cambridge: Cambridge University Press.

NISHIDA, T. (1987) – Local traditions and cultural transmission. In Smuts, B.; Cheney, D.; Seyfarth, R.; Wrangham; R. and Struhsaker, T., eds. – *Primate Societies*. Chicago: The University of Chicago Press, p. 462-474.

OAKLEY, K.P. (1965) – *Man the Tool-Maker*. London: British Museum (Natural History). 5th edition.

OHASHI, G. (2006) – Behavioural repertoire of tool use in the wild chimpanzees at Bossou. In Matsuzawa, T.; Tomonaga, M.; Tanaka, M., eds. – *Cognitive development in chimpanzees*. Tokyo: Springer, p. 439-451.

PERLÈS, C. (1992) – In search of lithic strategies a cognitive approach to prehistoric chipped stone assemblages In Gardin, J.; Peebles, C.S., eds. – *Presentations in Archaeology*. Bloomington IN: Indiana University Press, p. 223-247.

SAKURA, O.; MATSUZAWA, T. (1991) – Flexibility of wild chimpanzee nut-cracking behavior using stone hammers and anvils: an experimental analysis. *Ethology*. 87: p. 237-248.

SOUSA, C. (1999) – *The use of tokens by chimpanzees: A new approach to tool-use and intelligence*. Thesis for Master Degree in Human Evolution. Coimbra: University of Coimbra. Portugal.

WHITEN, A.; GOODALL, J.; MCGREW, W.C.; NISHIDA, T.; REYNOLDS, V.; SUGIYAMA, Y.; TUTIN, C.; WRANGHAM, R.; BOESCH, C. (1999) – Cultures in chimpanzees. *Nature*. 399: p. 682-685.

WHITEN, A.; BOESCH, C. (2001) – The cultures of chimpanzees. Scientific American 284 (1): p. 48-55.

WYNN, T.; MCGREW, W.C. (1989) – An ape's view of the Oldowan. *Man*. 24: p. 383-398.

VIEIRA, B. (1997) – Apresentação. *Ethnologia*. 5: p.11-14.

A ETOLOGIA E A ARTE PALEOLÍTICA: CERVÍDEOS E CAPRÍNEOS NA ARTE PALEOLÍTICA DO VALE DO CÔA

V.C.M. CARVALHO

Grupo de Estudos em Evolução Humana, Departamento de Antropologia, Universidade de Coimbra, Portugal

Abstract: *The interdisciplinary dialogue between ethology and the study of the palaeolithic art is recognised as fundamental, considering that, when regarding to its figurative representations, it is essentially a zoomorphic art. Concepts and methods of biology, zoology and ethology are combined, searching for an interdisciplinary analysis of the cervids and caprins palaeolithic representations from the Côa Valley.*

The animals knowledge and, particularly, of their behaviour allows a better understanding of the ways of life, the traditions and the manifestations of the human populations, especially when they have a crucial relation with their ecologic environment, as in the case of the Upper Palaeolithic human groups.

Key-Words: *Ethology, Palaeolithic Art, Côa Valley, Cervidae, Caprinae*

Résumé: *On considère essentiel le dialogue interdisciplinaire entre l'éthologie e l'étude de l'art paléolithique, une fois que cette est, dans son coté figurative, une art zoomorphique. Dans cette étude sont appliquées concepts et méthodes de la biologie, zoologie et éthologie pour une analyse interdisciplinaire des représentations des cervidés et caprinés dans l'art paléolithique de la vallée du Côa.*

La connaissance des animaux et en particulier de ses comportements permet une meilleure compréhension des moeurs de vie, coutumes et manifestations des populations humaines, essentiellement quand elles ont une relation important avec son milieu écologique, comme c'est le cas des groups humains du Paléolithique Supérieur.

Mots-clés: *Éthologie, Art Paléolithique, Vallée du Côa, Cervidés, Caprinés*

Resumo: *Considera-se fundamental o diálogo interdisciplinar entre a Etologia e o estudo da Arte Paleolítica, tendo em consideração que esta é, na sua vertente figurativa, uma arte essencialmente zoomórfica. Neste estudo combinam-se conceitos e métodos da biologia, zoologia e etologia com o objectivo de efectuar uma análise interdisciplinar das representações de cervídeos e caprídeos na arte paleolítica do Vale do Côa.*

O conhecimento dos animais e, em particular, do seu comportamento, permite uma melhor compreensão do modo de vida, costumes e manifestações das populações humanas, especialmente quando estas têm uma relação crucial com o seu meio ecológico, como é o caso dos grupos humanos do Paleolítico Superior.

Palavras-chave: *Etologia, Arte paleolítica, Vale do Côa, Cervídeos, Caprídeos*

A ARTE DO PALEOLÍTICO SUPERIOR E A ETOLOGIA

> "En fait, les chasseurs du paléolithique supérieur ont été les premiers "éthologistes", et l'attitude éthologique naît avec leurs capacités d'observation et figuration de ce qu'ils ont observé sur le terrain, en toute objectivité." (Vieira, 1995: 579)

O Paleolítico Superior corresponde a um estimulante período de mudanças, de dinamismo cultural e demográfico, propício a modificações comportamentais. Neste contexto cultural, e após uma longa fase em que se pode considerar que existem sinais de uma emergência da arte, durante o Paleolítico Médio e nas culturas de transição, assiste-se a uma eclosão artística, entre 36.000 e 30.000 BP, surgindo a arte parietal e a arte figurativa. O início do Paleolítico Superior corresponde a uma mudança quantitativa e qualitativa, marcada por um acréscimo de indícios de comportamento simbólico, tais como a proliferação de marcas e traços geométricos sobre osso e pedra, o aparecimento da arte rupestre figurativa, a multiplicação de adornos e de sepulturas. Estes indícios podem ser interpretados como resultantes de um desenvolvimento das actividades rituais e das crenças relativas ao destino humano (Lorblanchet, 1995; Djindjian *et al.*, 1999; Lorblanchet, 1999).

A adopção do conceito de ***arte*** aplicado a um vasto conjunto de formas, que testemunham comportamentos simbólicos e sociais dos caçadores paleolíticos, tem vindo a suscitar um intenso debate científico. O conceito de ***arte*** aplica-se neste caso à presença regular de grafismos figurativos ou não, repetidos sobre paredes rochosas ou sobre objectos, e constituindo um fenómeno cultural estruturado. O aparecimento da arte rupestre, parietal ou móvel, não corresponde a um evento homogéneo, sincrónico e universal. O nascimento da arte pode ser encarado como uma resposta a necessidades variadas, resultando de crenças particulares, de escolhas culturais e simbólicas, e valorizando técnicas, temas e suportes materiais diferentes (Lorblanchet, 1999).

A noção de que a Etologia, a Biologia ou a Zoologia podem dar um importante contributo na análise e interpretação das representações figurativas dos povos do Paleolítico Superior não é nova, contudo, o estabelecimento de uma relação coerente entre estes campos tem-se revelado complexo e algo controverso (Bandi *et. al.* 1984; Bandi, 1984; Vieira, 1995; Paillet, 1999; Apellániz, 2001;

Clottes & Azema, 2005; Gely & Azema, 2005; Guthrie, 2005; Bahn, 2006).

A arte figurativa paleolítica é fundamentalmente zoomórfica, sendo o número de representações de animais bastante mais numeroso que o de humanos, quer na arte parietal quer na arte móvel (Leroi-Gourhan, 1992). A lista convencional de animais representados no Paleolítico Superior engloba a maior parte dos mamíferos visíveis nos ecossistemas destes grupos humanos. Pelo contrário, apenas se encontra figurado um número reduzido de espécies de aves, de répteis, de anfíbios e de peixes (Lorblanchet, 1995:49). Deve-se salientar que de suporte para suporte, de região para região, de sítio para sítio e mesmo de fase para fase, as espécies reproduzidas, bem como a sua representatividade, são distintas. Parece existir uma concordância entre as espécies de animais representadas e as espécies existentes nos meios ecológicos, habitados pelos grupos humanos, possivelmente, responsáveis por essas mesmas obras. Por outro lado, parece não existir uma relação equitativa entre as espécies consumidas, ou que compunham o meio ecológico, e as espécies representadas (Otte, 1999).

André Leroi-Gourhan (1990; 1992) questionava-se se os animais pré-históricos constituíam uma espécie de panteão ou apenas uma simples "despensa" e provisão de alimentos, salientando a lista restrita das espécies representadas. Leroi-Gourhan (1992) concluía que mais do que um catálogo cinegético, as figuras animais constituíam um "bestiário", ou seja uma lista convencional de animais que participavam numa certa mitologia.

Delporte afirma que as relações entre o Homem e os animais seriam diferentes das que existiam entre o homem e os objectos que este manipulava – "o animal era algo mais que um simples "objecto para matar e comer"; o homem devia "pensar" nas reacções e na maneira de viver do animal e, também nas suas migrações". O vínculo entre o homem e o animal escapar-se-ia assim do mundo material (Delporte, 1995:11). Este investigador considera ser fundamental a compreensão das relações existentes entre o mundo animal e os grupos humanos do Paleolítico Superior, uma vez que aquelas se enquadram no seio de uma economia de subsistência na qual a caça, a pesca e a recolecção condicionavam fortemente a vida humana (Delporte, 1995:210).

Geist (1978:322) ao referir-se à arte paleolítica afirma que esta "(...) reveals a superb power of observation and exactness of mental notation, and therefore a high degree of familiarity with the animals they painted."

A relação entre a etologia e a arte paleolítica foi abordada por Vieira (1995) que apresenta alguns casos de representações paleolíticas de animais em que os comportamentos típicos da espécie são alvo de uma figuração muito rigorosa, e que podem ser, segundo o autor, analisados no quadro de uma etologia objectiva. Este autor sublinha que na arte paleolítica, por vezes, se podem identificar episódios de corte e de combate ritualizado no contexto da selecção sexual, figurados em pormenor e com bastante rigor (Vieira, 1995). O autor considera que a maior parte dos comportamentos passíveis de serem reconhecidos nas espécies figuradas são comportamentos sexuais ou agonísticos em contexto sexual: marcação territorial, solicitação sexual traduzida em sinais de chamamento durante a corte, de aproximação inter-sexual, de combates rituais intra-sexuais entre machos rivais (Vieira, 1995:585).

Na perspectiva de Cremades (1997) as representações paleolíticas de animais podem, por vezes, ser indicativas de sazonalidade, representando determinadas época do ano. A autora baseia-se em dados da zoologia e da etologia, analisando múltiplos componentes tais como as características da pelagem, os ciclos de reprodução, o crescimento das hastes, os ciclos migratórios, e alguns comportamentos sazonais distintivos.

A percepção de que o conhecimento do meio biológico, da morfologia e da etologia animal é fundamental para um estudo aprofundado das representações gráficas dos animais da arte paleolítica, encontra-se subjacente à análise efectuada por Iakovleva e Pinçon (1998) sobre as esculturas de cabra-montês do sítio de Angles-sur-L'Anglin (França). As investigadoras basearam-se em dados morfométricos e morfológicos ao compararem as representações gráficas com as descrições etológicas relativas à espécie actual de *Capra ibex,* procurando analisar elementos indicadores de diagnose sexual, dimorfismo sexual, idade, e eventuais interacções sociais entre os indivíduos. As conclusões deste estudo apontam para a existência, ao nível das representações gráficas, de indivíduos, de ambos os sexos e de várias classes etárias, esculpidos de forma proporcional às dimensões reais da espécie, e representados em cenas que parecem ser concordantes com a organização social da espécie. Após uma descrição detalhada de cada figura, e considerando a análise de conjunto, é colocada a hipótese de estarmos perante uma representação geral de um movimento sazonal: a migração de Outono.

Nos últimos anos foram publicados alguns trabalhos (Paillet, 1999; Clottes & Azema, 2005; Gely & Azema, 2005) que procuraram estabelecer uma relação mais estreita entre a biologia e o estudo da arte dos grupos humanos do Paleolítico Superior. Estas investigações, focalizadas em representações de determinados géneros ou espécies de animais, numa dada região ou sítio arqueológico, constituem exemplos particularmente interessantes das potencialidades e dos limites da aplicação deste conceito de análise interdisciplinar.

A ARTE DO PALEOLÍTICO SUPERIOR DO VALE DO CÔA

A arte paleolítica do Vale do Côa foi identificada em finais de 1991, aquando da descoberta da rocha 1 da

Canada do Inferno, contudo, apenas em finais de 1994 se tornou pública a sua descoberta (Zilhão, 1997b). Esta revelação desencadeou um processo polémico uma vez que as gravuras identificadas se encontram numa zona que seria inundada caso a construção da barragem no rio Côa, já iniciada à época, fosse concluída. A preservação e o estudo da arte rupestre foram assegurados em finais de 1995, com o abandono do projecto da EDP e com a posterior criação do Parque Arqueológico do Vale do Côa e do Centro Nacional de Arte Rupestre. A importância deste património foi reconhecida em 1998 pela UNESCO ao classificar o Vale do Côa como Património da Humanidade.

Uma das principais implicações desta descoberta foi a da consciencialização da importância da arte rupestre ao ar livre, no contexto da arte do Paleolítico Superior Europeu. No Vale do Côa encontra-se "o maior ciclo temporal de arte rupestre da Europa Ocidental" (Baptista, 1999:20) que compreende representações desde o Paleolítico Superior até aos nossos dias.

Os dados publicados referem 28 sítios com arte rupestre, identificados ao longo do Vale do Côa, desde o sítio da Faia até à foz, bem como em pequenos vales subsidiários e na área do rio Douro. A arte paleolítica encontra-se presente em 24 destas estações arqueológicas e é, numericamente, a mais significativa (Baptista, 1999; Baptista, 2001).

As representações têm como suportes preferenciais painéis em xisto de grande dimensão com superfícies orientadas verticalmente. Na sua maioria, as representações identificadas, e que persistiram até aos nossos dias, foram elaboradas com recurso à técnica da gravura. A arte paleolítica do Côa tem uma temática essencialmente zoomórfica, e apresenta um bestiário pouco variado, composto, segundo Baptista (1999:30), genericamente, por capríneos, equídeos, bovídeos e cervídeos, com presença indiscriminada de machos e fêmeas. Os animais indeterminados, ou seja "(...) com caracteres morfológicos pouco explícitos ou insuficientes para se atribuírem com segurança a uma determinada espécie (Baptista, 1999:31)", constituem uma percentagem elevada. A fauna representada no Côa corresponde, essencialmente, aos herbívoros de grande porte mais característicos dos ecossistemas da Meseta Ibérica no Paleolítico Superior (Zilhão, 1997a; Zilhão 1997b; Baptista, 1999).

Os motivos, essencialmente zoomórficos, com presença pouco regular de signos e muito escassa de representações humanas, são similares aos que se detectam na arte parietal da Península Ibérica e na região Franco – Cantábrica, encontrando-se no Côa algumas convenções estilísticas que estão igualmente presentes nestas regiões. No entanto, algumas das características mais notáveis do Côa são menos comuns como: a sobreposição e acumulação intencional de motivos em zonas circunscritas dos painéis ou a representação de múltiplas cabeças ou patas, sendo esta multiplicação de elementos anatómicos interpretada como passível de representar movimento (Baptista, 2001).

Os dados obtidos nos sítios arqueológicos intervencionados apontam para uma ocupação humana do vale durante vários períodos do Paleolítico Superior, designadamente, no Gravettense, Solutrense e Magdalenense (Aubry, 2002). As escavações realizadas, nomeadamente, na estação arqueológica do Fariseu, têm permitido ainda a identificação de peças de arte móvel, de cronologia paleolítica (Garcia Díez & Aubry, 2002).

Baptista (1999:33) considera que "as características estilísticas desta gramática figurativa (...) conduzem-nos (...) para dois grandes momentos, evoluindo a grande generalidade do ciclo quaternário do Côa pelo menos entre o Gravettense e o Magdalenense Final".

OBJECTIVOS E MÉTODOS

O principal objectivo deste estudo consiste em detectar novos indicadores relativos à percepção que as populações de caçadores-recolectores do Paleolítico Superior do Vale do Côa possuíam sobre os animais que representavam e sobre o seu comportamento, a partir da análise das suas representações de cervídeos e capríneos.

Este propósito pressupõe uma abordagem a duas disciplinas aparentemente muito distintas, o estudo da arte paleolítica e a etologia, de modo a discutir o conhecimento que os grupos humanos do Paleolítico Superior tinham do comportamento de determinadas espécies animais, designadamente as que caçavam e representavam, verificando se o comportamento das espécies animais seria efectivamente plasmado na arte dos que o observavam.

A proposta que se apresenta incide sobre uma amostra previamente definida, que corresponde às figuras, interpretadas como representações de cervídeos e capríneos, e que se encontram publicadas pela equipa responsável pelo estudo da arte do Côa. Efectua-se uma análise individual de cada representação gráfica, sendo consideradas as mesmas variáveis para cada figura. A análise das representações é efectuada através de observações múltiplas das representações originais *in situ* e de fotografias e decalques das mesmas. Os dados da análise das representações de cervídeos e capríneos na arte paleolítica do Vale do Côa são comparados com dados provenientes de estudos específicos no âmbito da etologia, relativos a essas espécies (Clutton-Brock *et al.*, 1982; Clutton-Brock *et al.*, 1988; Geist, 1999; Sáenz de Buruaga *et al.*, 2001; Fonseca, 2003; Alcock, 2005). Procura-se comparar os dados relativos às espécies representadas relativamente às consumidas, considerando os dados arqueológicos da região para o Paleolítico Superior.

Uma vez que considerámos necessário um período de familiarização com as espécies encontra-se em curso uma observação *ad libitum* (Martin & Bateson, 1993) de animais em *habitat* natural, na serra da Lousã, de duas espécies consideradas na amostra, o veado (*Cervus elaphus*) e o corço (*Capreolus capreolus*).

A investigação parte do pressuposto, para efeitos teóricos, que as espécies actualmente observáveis têm comportamentos semelhantes às espécies observadas pelos grupos humanos à época de realização das representações gráficas.

Este estudo realiza-se em estreita colaboração com o Centro Nacional de Arte Rupestre, tendo o investigador Martinho Baptista autorizado o estudo da amostra em análise e disponibilizado para tal a base de dados gráficos – fotografia e decalque. O Parque Arqueológico do Vale do Côa permitiu a realização dos trabalhos de campo e autorizou e apoiou logisticamente as visitas necessárias aos núcleos de arte do Vale do Côa. A observação *ad libitum* de Veado e Corço, na Serra da Lousã, só é possível graças ao apoio prestado por Carlos Fonseca e a sua equipa, do Departamento de Biologia da Universidade de Aveiro.

A APLICAÇÃO DO CONCEITO: UM CASO "TIPO"

No âmbito da representação artística de padrões de comportamento animal, o caso das gravuras interpretadas como veados machos parece constituir um bom exemplo da aplicação deste conceito de abordagem interdisciplinar.

Considerando algumas características especificas na análise das figuras, como a morfologia, a postura e o comportamento observável, podemos colocar a hipótese de que algumas gravuras paleolíticas do Vale do Côa correspondem à representação objectiva de indivíduos adultos, possivelmente, da espécie *Cervus elaphus,* do sexo masculino, durante o período da brama e, em determinados casos, numa postura que indicia um comportamento de vocalização (Clutton-Brock *et al*., 1988; Geist, 1999; Sáenz de Buruaga *et al*., 2001; Fonseca, 2003; Alcock, 2005). As seguintes figuras constituem bons exemplos deste caso "tipo" de gravuras: o veado da Rocha 11B da Canada do Inferno, o veado raspado da Rocha 10C da Penascosa e o veado da Rocha 2 da Penascosa (vide fig. 19.1). Estas figuras integram, igualmente, o grupo de gravuras consideradas como indicativas da representação das variações sazonais na arte paleolítica, uma vez que este comportamento se conota com a época de acasalamento desta espécie. A postura em que, nestes casos, são representados os animais são indicativas de uma figuração de movimento relacionada com um comportamento de vocalização, de carácter sexual e agonístico.

A competição entre machos pode ser observada nos veados adultos da espécie *Cervus elaphus,* uma espécie amplamente representada e consumida no Paleolítico Superior. As hastes dos veados machos adultos crescem cada ano, e na época de acasalamento, no Outono, eles confrontam-se directamente pelo acesso às fêmeas, que não possuem hastes e que são agrupadas por sucessivos machos em haréns. Os machos desafiam-se inicialmente recorrendo a bramidos, o que lhes permite avaliarem-se mutuamente. Este processo de avaliação mútua pode culminar, após uma fase de aproximação dos machos rivais, num confronto físico directo. Os machos entrelaçam as hastes e empurram-se um contra o outro, vencendo habitualmente o animal mais corpulento (Clutton-Brock *et al.,* 1982; Clutton-Brock *et al.,* 1988).

PERSPECTIVAS FUTURAS

A ideia de estabelecer uma ponte entre a Etologia e o estudo da arte dos caçadores do Paleolítico Superior europeu parece-nos pertinente considerando que esta é, na sua vertente figurativa, uma arte essencialmente zoomórfica. A etologia, na sua busca pela compreensão do comportamento animal, pode dar neste campo de estudos um contributo muito importante no âmbito de um verdadeiro diálogo interdisciplinar.

Consideramos que o conhecimento do comportamento dos animais e da sua morfologia permite uma melhor compreensão do modo de vida, costumes e manifestações das populações humanas, especialmente, quando estas têm uma relação crucial com o seu meio ecológico, como é o caso das populações de caçadores-recolectores do Paleolítico Superior.

As relações existentes entre os grupos humanos do Paleolítico Superior e os restantes animais podem ser percepcionadas, se bem que de forma parcelar, com base na análise de contributos indirectos, obtidos a partir das evidências materiais, como os utensílios líticos e ósseos, e dos vestígios de caça, desmanche, preparação e consumo, aos quais se somam aqueles que nos mostra a arte.

As variadas interpretações propostas para estas expressões gráficas não eliminam a complexidade da análise do papel dos animais nas pinturas e gravuras das populações humanas do paleolítico superior. Consideramos que estas representações devem reflectir não apenas o conhecimento que as populações teriam do seu mundo real, mas também todo um sistema mítico e simbólico.

As representações de animais na arte do Paleolítico Superior europeu, interpretadas genericamente como *naturalistas*, e que se considera serem as que de forma mais sustentada podem permitir o estabelecimento de paralelos com o estudo do comportamento animal, apesar de serem controversas, parecem-nos, de facto, constituir, pelo menos em alguns casos, uma base séria para uma leitura comparativa entre as figuras e o comportamento das espécies que estas representam, tendo sempre presente o necessário espírito analítico e crítico.

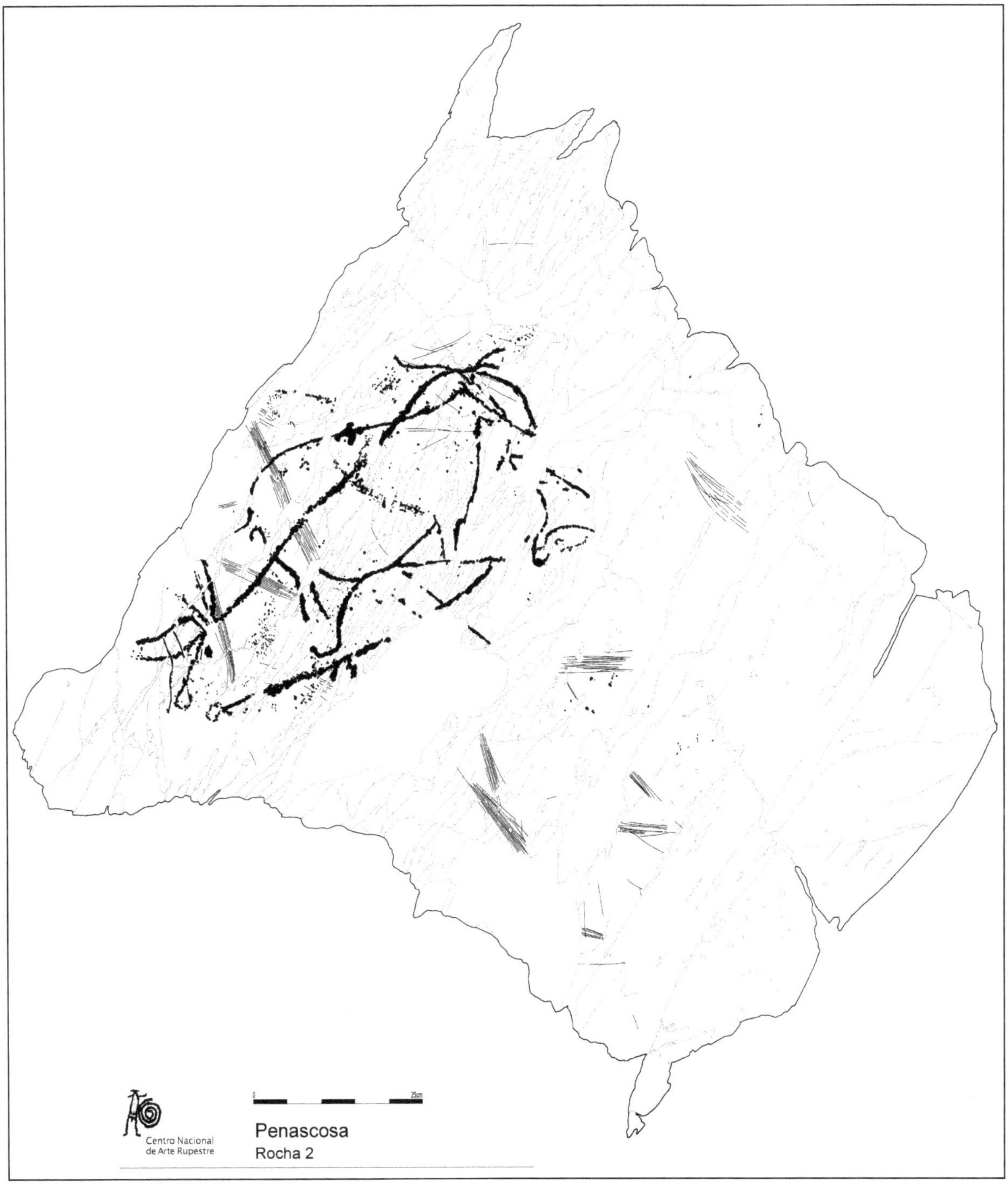

Fig. 19.1. Veado, virado para a esquerda, da Rocha 2 da Penascosa, com a boca aberta e numa postura que indicia um comportamento de brama

O estudo, que se encontra em curso, pretende efectuar uma análise objectiva deste tipo de representações gráficas, enquadrada no âmbito de uma abordagem interdisciplinar que recorre a conceitos e conhecimentos diversos provenientes da etologia, da zoologia, da antropologia, da arqueologia pré-histórica e da arte paleolítica, e trazer, deste modo, resultados que abram novas perspectivas ao conhecimento sobre as múltiplas interacções Homem-Animal, especificamente no Paleolítico Superior do Vale do Côa.

Bibliografia

ALCOCK, J. (2005) – *Animal behavior: an evolutionary approach*. Eight Edition. Massachussetts (EUA): Sinauer Associates, Inc.

APELLÁNIZ, J.M. (2001) – *La abstracción en el arte figurativo del Paleolítico*. Bilbao: Universidad de Deusto.

AUBRY, T. (2002) – Le contexte archéologique de l'art paléolithique à l'air libre de la vallée du Côa. In

Sacchi, D., eds. – *L'art paléolithique à l'air libre: le paysage modifié par l'image (Tautavel, Campôme, 7-9 octobre 1999)*. Saint-Estève: GAEP, GÉOPRE, p. 25-38.

BAHN, P. (2006) – Sex and violence in Rock art. *Nature*. 441: p. 575-576.

BANDI, H.-G.; HUBER, W.; SAUTER, M.; SETTER, B. (1984) – *La contribution de la zoologie et de l' éthologie à l' interprétation de l'art des peuples chasseurs préhistoriques*. 3ª Colloque de la Société Suisse des Sciences Humaines – 1979. Fribourg: Editions Universitaires Fribourg Suisse.

BANDI, H.-G. (1984) – Contribution de la zoologie et de l'éthologie a l'interprétation de l'art des peuples chasseurs préhistoriques. *L'Anthropologie*. Tome 88: 4. p. 563-571.

BAPTISTA, A. (1999) – *No tempo sem tempo. A arte dos caçadores paleolíticos do Vale do Côa, com uma perspectiva dos ciclos rupestres pós-glaciares*. Vila Nova de Foz Côa: Parque Arqueológico Vale do Côa.

BAPTISTA, A.M. (2001) – The Quaternary Rock Art of the Côa Valley. In Zilhão, J.; Aubry, T.; Carvalho, A.F., eds. – *Les premiers hommes modernes de la Péninsule Ibérique (Actes du Colloque de la Commission VIII de l'UISPP. Vila Nova de Foz Côa, 22-24 Octobre 1998)*. Lisboa: IPA, p. 237-252.

CLOTTES, J.; AZEMA, M. (2005) – *Les Félins de la grotte Chauvet*. Paris: Seuil.

CLUTTON-BROCK, T.S.; GUINNESS, F.E.; ALBON, S.D. (1982) – *Red deer: behavior and ecology of two sexes*. Chicago: Wildlife Behavior and Ecology Series.

CLUTTON-BROCK, T.; ALBON, S.; GUINNESS, F. (1988) – Reproductive success in male and female Red Deer. In Clutton-Brock, T., eds. *Reproductive success. Studies of individual variation in contrasting breeding systems*. Chicago: The University of Chicago Press, p. 325-343.

CREMADES, M. (1997) – La representation des variations saisonnières dans l'art paleolihique. *L'Anthropologie*. 101:1, p. 36-82.

DELPORTE, H. (1995) – *La imagen de los animales en el arte prehistorico*. Madrid: Compañia Literária.

DJINDJIAN, F., KOSLOWSKI, J., OTTE, M. (1999) – *Le paléolithique supérieur en Europe*. Paris: Armand Colin.

FONSECA, C. (2003) – *Veados (Cervus elaphus L.) na Serra da Lousã*. Aveiro: Departamento de Biologia da Universidade de Aveiro.

GARCÍA DÍEZ, M.; AUBRY, T. (2002) – Grafismo mueble en el Valle del Côa (Vila Nova de Foz Côa, Portugal): la estación arqueológica de Fariseu. *Zephyrus*. Salamanca. 55: p. 157-182.

GEIST, V. (1978) – *Life strategies, human evolution, environmental design. Toward a biological theory of health*. New York: Springer-Verlag.

GEIST, V. (1999) – *Deer of the World: their evolution behaviour and ecology*. Shrewsbury: Swan-Hill Press.

GELY, B.; AZEMA, M. (2005) – *Les Mammouths de la grotte Chauvet*. Paris: Seuil.

GUTHRIE, R. (2005) – *The nature of paleolithic art*. Chicago: The University of Chicago Press.

IAKOLVLEVA, L.; PINÇON, G. (1998) – The upper palaeolithic sculptured ibex of Angles-sur-L'Anglin, France. *Oxford Journal of Archaeology*. 17: 3, p.257-268.

LEROI-GOURHAN, A. (1990) – *As religiões da Pré-história*. Lisboa: Edições 70.

LEROI-GOURHAN, A. (1992) – L'*art parietal. Langage de la préhistoire*. Grenoble: J. Million.

LORBLANCHET, M. (1995) – *Les grottes ornées de la préhistoire. Nouveaux regards*. Paris: Errance.

LORBLANCHET, M. (1999) – *La naissance de l'art: Genèse de l'art préhistorique*. Paris: Errance.

MARTIN, P.; BATESON, P. (1993) – *Measuring behaviour: an introductory guide*. Second edition. Cambridge: Cambridge University Press.

OTTE, M. (1999) – *La Préhistoire*, Paris: De Boeck & Larcier.

PAILLET, P. (1999) – *Le bison dans les arts magdaléniens du Périgord*. XXXIII supplément à Gallia Préhistoire. Paris: CNRS éditions.

SÁENZ DE BURUAGA, M.; LÚCIO CALERO, A.; PURROY IRAIZOZ, F. (2001) – *Reconocimiento de sexo y edad en especies cinegéticas*. León: Edilesa.

VIEIRA, A.B. (1995b) – Art paléolithique et éthologie. *Trabalhos de Antropologia e Etnologia*. XXXV (Fasc.4). Porto.

ZILHÃO, J. (1997a) – *O Paleolítico Superior da Estremadura Portuguesa*. Lisboa: Edições Colibri. Vol. I e II.

ZILHÃO, J. (Coord.) (1997 b) – *Arte rupestre e pré-história do Vale do Côa. Trabalhos de 1995-1996. Relatório científico ao governo da República Portuguesa elaborado nos termos da resolução do conselho de Ministros nº4/96 de 17 de Janeiro*. Lisboa: Ministério da Cultura.

MARKERS OF OCCUPATIONAL STRESS: INTERDISCIPLINARY APPROACH

Sandra ASSIS

Grupo de Estudos em Evolução Humana (GEEvH), Department of Anthropology, Faculty of Sciences and Technology, University of Coimbra, email: sandraassis78@yahoo.com

***Abstract**: The human skeleton has the ability to show many aspects of past populations life, especially the traces related with their everyday activities and occupations.*

In this field, a considerable number of paleopathological studies have pointed out a straight connection between some particular types of bone lesions and the stress applied during the performance of certain daily activities. This issue introduces the main goal of this research. Based on a skeletal sample (N=64) exhumed from the necropolis of Constância (XIV-XIX centuries), an interdisciplinary analyis will be developed which combines the biological information from the skeleton, through the study of markers of occupational stress (MOS), with the historic, ethnographic and economic factors. This documental sources shows a significant human dependence from the rivers activities, namely fishing, ship building, transport of goods and people.

***Key-words**: Constância, rivers, interdisciplinary approach, occupational patterns, MOS*

***Résumé**: Le squelette humain a la capacité d'enregistrer beaucoup d'évènements qui restent comme des marques, imprimées par la répétition des taches quotidiennes.*

A ce propos plusieurs études paléopathologiques ont établi la relation entre certaines lésions aux os et le "stress" inhérent a l'exécution des activités quotidiennes.

Voici le but très précis de cette investigation: En partant des squelettes trouvés a l'ancienne nécropole de Constância, (siècle XIV-XIX), on essayera de mener une abordage interdisciplinaire par la combinaison des connaissances au niveau de la biologie du squelette, l'étude des marqueurs de stress occupationnel, et les données historiques, ethnographiques et économiques, (attribuant aux fleuves et aux activités qu'y sont associés, comme la pèche, la construction des bateaux, le transport des marchandises et des personnes), le statut d'activités principales entre les anciens habitants de Constância.

***Mots-clés**: Constância, fleuves, abordage interdisciplinaire, normes professionnelles, MSO*

***Sumário**: O esqueleto humano tem o potencial de registar muitos aspectos da vida das populações do passado, especialmente, os traços envolvidos nas suas actividades quotidianas e ocupacionais.*

Neste campo vários estudos paleopatológicos têm revelado alguma correlação entre alguns tipos de lesões ósseas e o stresse imposto durante a realização de determinadas actividades. Este intento introduz o objectivo fulcral da investigação. A partir de uma amostra de esqueletos (N=64) exumados da antiga necrópole de Constância (séculos XIV-XIX), conduzir uma abordagem interdisciplinar, combinando a informação da biologia do esqueleto, através do estudo de marcadores de stresse ocupacional (MSO), com dados históricos, etnográficos e económicos que atribuem aos rios e às actividades que deles derivaram, nomeadamente a pesca, a construção naval e o transporte de mercadorias e pessoas, enquanto ocupações dominantes entre os antigos habitantes de Constância.

***Palavras-chave**: Constância, rios, abordagem interdisciplinar, padrões ocupacionais, MSO*

INTRODUCTION

The knowledge of past populations has motivated, since early periods, the collection of data in order to enable the reconstruction of their *modus vivendi*. During centuries, this exercise was limited to gathering fragments of material culture; however, a new edifying source has lately enriched the cognitive field – the human skeleton.

The human skeleton allows the building of a bridge with past lives, in the way that it can record important episodes of the individuals' ontogenetic courses, associated with their diet, health conditions, behaviour and occupation (Larsen, 2000). In this field, several studies have revealed a relationship between some types of lesions and the stress applied to the bones during the performance of certain activities. This assumption is based on the principle that the bone can react to biomechanical forces induced by quotidian actions, developing pathological conditions like osteoarthrosis, enthesopathies and trauma, and/or non-pathological as accessory facets or hypertrophy of muscle-skeletal insertions (Larsen, 1997; Steinbock, 1976; Merbs, 1989; Kennedy, 1998 *in* Rodrigues Carvalho, 2004). Through the study of the above mentioned lesions, identified as markers of occupational stress (MOS), we can potentially explain some aspects connected with the social organization, activity and economy of the ancient groups, thus opening a window to the past.

MARKERS OF OCCUPATIONAL STRESS (MOS)

Presentation note

The study of activity, and respective bone manifestations, is a subject with a long history and it has been perpetuated by different branches of scientific knowledge (Robb, 1994).

In fact, a wide range of disciplines have focused on the thematic of markers of occupational stress (MOS), namely the Occupational and Sports Medicine, the Forensic Anthropology and the Paleoanthropology.

The Occupational and Sports Medicine studies the work conditions and the subsequent stress factors that predispose workers and athletes to bone lesions, including degenerative joint diseases. Recent data presented by the World Health Organization (WHO) indicates that in Canada, Finland and the United States more people are disabled from working as a result of musculoskeletal disorders (37%) – especially back pain – than from any other group of diseases, like noise (hearing loss) 16% or risk of injuries (8%) (Nelson *et al.*, 2005).

In the field of Forensic Anthropology, the analysis of MOS is used to identify particular characteristics of the skeleton that may be related to occupational demands in which an individual might have be engaged during life, to permit its positive identification (Kennedy, 1989).

To Paleoanthropology, the study of occupation represents an important instrument to access and reconstruct past populations lives. Through the MOS it is possible to understand some aspects linked to material culture, economy, ecology, social organization, sexual labour division and general health condition (Kennedy, 1989; Kumar, 2001). In sum, it's a way to understand the relationship between biology and human culture.

Concepts and definitions

The Markers of Occupational Stress (MOS) represent a bone response to the customary activity and included osteoarthrosis, musculoskeletal markers (MSM), postural markers and stress fractures.

Osteoarthrosis (OA) is a degenerative bone disease characterized by subchondral sclerosis, eburnation, lipping and porosity of articular surfaces (Rogers *et al.*, 1987 *in* Wilczak and Kennedy, 1998; Queiroz, 2000; Cunha, 2003). Now, like in the past, OA is the most common condition observed in the osteoarchaeological record and one of the main indicators of occupational stress (Roberts and Manchester, 2005).

The musculoskeletal markers (MSM) are distinctive skeletal traces, occurring where a muscle, tendon or ligament inserts on the bone surface after a recurrent use (Hawkey and Merbs, 1995; Steen and Lane, 1998). This skeletal manifestation can be expressed through bone hypertrophy or, in severe cases, through the formation of enthesopathies, "(...) a morbid condition, or disease at the site of attachment of a muscle or ligament to bone (enthese)" (Anderson, 1994 *in* Steen and Lane, 1998: 341).

The stress fractures represent another source of information. These are produced during the performance of continuous movements which overpass the normal resistance of the bone. Furthermore, they represent a structural failure to abnormal biomechanical stress (Martin & McCulloch, 1985 *in* Iwamoto & Takeda, 2003; Daffner, 1978 *in* Niva, 2006).

In agreement with Rodrigues-Carvalho (2004), the postural markers denote a skeleton response to the adoption of certain corporal positions. In this domain, the most frequent observed traces are the squatting facets of tibiae and talus (Mays, 1998).

A historical review

The scientific interest on the study of occupation and its effects on the health of people is not recent, dating from the XVI century. In fact, Georgius Agricola (1494-1555) and Paracelsus (1493-1541) were the first physicians to study the effect of polluted air in the health of miner workers. Lately, in the XIX century, William Lane (1856 – 1943), an anatomist, used the expression "pressure changes" (1887) to refer the anatomical markers developed in response to habitual activities: a prelude to the future "markers of occupational stress" (Kennedy, 1989). Paul Broca (1824 – 1880) and Pierre Manouvrier (1850 – 1927), correlated platycnemia of the tibiae with the hyperactivity of muscles linked to locomotion. In 1892, the anatomist Julius Wolff (1836 – 1902) presented the Law of Transformation, named as Wolff's Law, in what: *"the form of the bone being given, the bone elements place or displace themselves in the direction of the functional pressure and increase or decrease their mass to reflect the amount of functional pressure"* (Kennedy, 1989: 134).

Other studies can be referred, as the research of Lawrence Angel (1946) based on the correlation of skeletal changes of the pelvis and the lower extremities of the ancient Greeks, during Neolithic, with locomotor's stress; or the analysis of Trinkaus (1975) concerning the study of squatting facets among Neanderthals (Kennedy, 1989). The last decades were proficient in new approaches and methods. Hawkey and Merbs (1995) created a method to study and classify MSM among the Ancient Hudson Bay Eskimos, refined by Mariotti *et al.* (2004). In 1999, Capasso and colleagues presented an "Atlas of Occupational Markers on Human Remains" that represents a reference for the analysis and interpretation of bone conditions. Recently, the new approaches have focused on the bone dynamic and its geometric proprieties in response to exercise.

In Portugal few studies have been conducted in this area, however, we can refer the research of Cunha and Umbelino (1995) concerning the analysis of enthesopathies in the Collection of Identified Skeletons from the Anthropological Museum of Coimbra University; the research of Cardoso and Lopes (2002) based on a skeleton discovered in Alcabideche (Cascais), and the study of MSM in a sample of skeletons recovered from São Miguel de Odrinhas (Ferreira, 2006). Finally, it's pertinent to refer the PhD project of Cardoso (w.d.) by the Durham University intituled: "A portrait of Gender in Two 19th/20th Portuguese Populations: A paleopatho-logical perspective".

Methodological limitations

The MOS study is very useful regarding the knowledge of the behaviours of past populations, and especially concerning the need to fulfil the time gap, being able to reconstruct their occupation and life styles. Even so, it is possible identify some methodological limitations, correlated at some point with the nature of the bone.

The bone tissue it's characterized by some monotony in its response to adverse factors. In MOS analysis, this propriety is a limitative factor for the interpretation of lesions. An example is provided by osteoarthrosis. It is frequent to attribute the aetiology of OA to biomechanical factors. However, and for the past populations, there is a lack of information about the stress factors that conditioned their existence, namely its severity, anatomical distribution and duration. In agreement with Wilczak and Kennedy (1998), for a correct interpretation of OA is necessary to evaluate the articular surfaces in detriment of the peripheral areas, predominantly affected by age; and to observe the main articulations engaged in the body movements, as the elbow and knee; and to record its asymmetrical distribution.

In the particular case of MSM, researchers have identified five major problems:

1) Muscle insertion sites are morphologically complex and difficult to observe consistently (Robb, 1998);
2) Different muscle are needed in order to perform body movements (Stirland, 1998);
3) Muscle groups respond to a complex mosaic of biomechanical stress (Robb, 1998);
4) Difficulty in distinguishing lesions that can be related to age changes, trauma or habitual activities (Robb, 1998);
5) We are limited by our knowledge and imagination to choose between the ranges of possible tasks performed by ancient populations (Robb, 1998).

The solution of these problems can be found through an interdisciplinary approach, which attempts to unify the biological information of the skeleton, throughout the paleopathological analysis of MOS, combined with the historical and ethnographical memory of ancient people. This integrative view is actually recognized as an important method to understand the evolution of scientific knowledge and it gives a new insight, by looking to humans as a complex bio-socio-cultural unity.

A CASE STUDY FROM CONSTÂNCIA

Constância is a small village located in the Santarém District (Centre of Portugal), edified in the confluence between the Tagus and Zêzere rivers (figure 20.1).

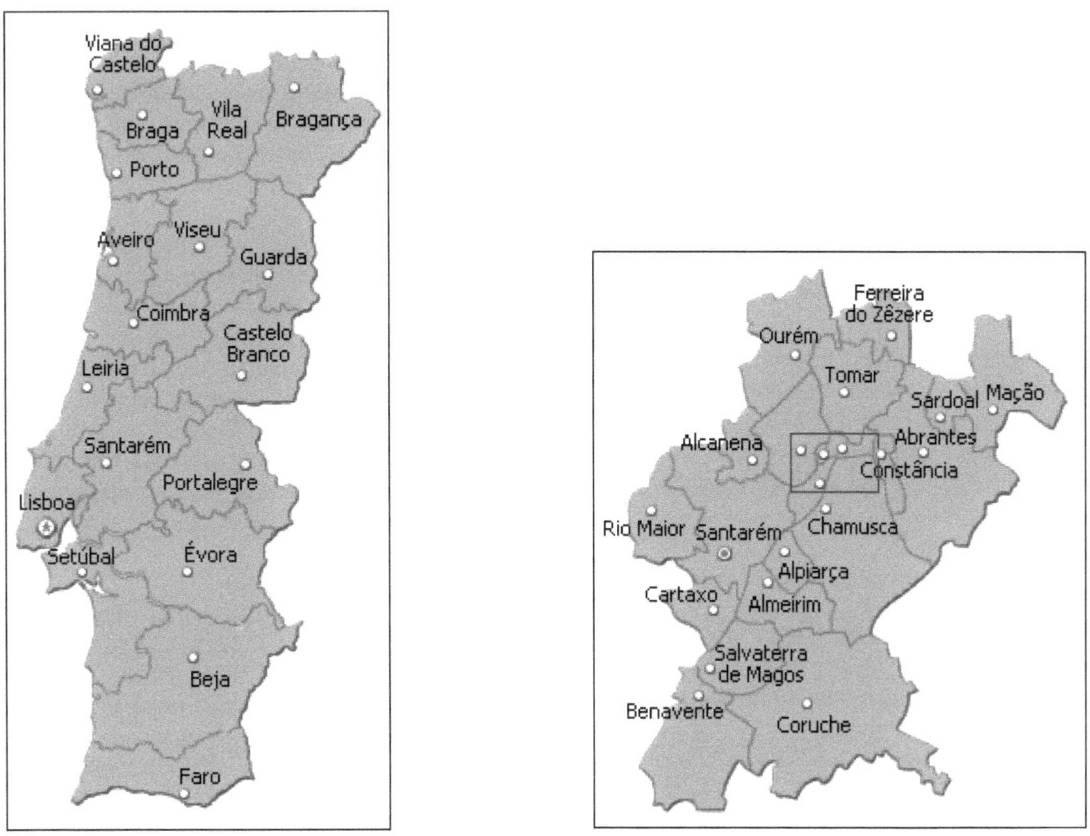

Fig. 20.1. Geographic location of Constância (adapted from http: // Portugal. veraki. pt)

Fig. 20.2. The village of Constância with some human activity in the Tagus River (adapted from Coelho, 1998: 15)

This proximity with two rivers was crucial to the economy of Constância, a reality that survived until the end of the XIX century (figure 20.2). Besides the economy, this close relationship with the water element may have been recorded in the bones of their ancient inhabitants, as a reflex of human occupation.

Goals

Having in consideration an interdisciplinary approach, the main goals of the project will focus on:

1) The activity patterns of ancient people from Constância based on the examination of degenerative, musculoskeletal, traumatic and postural bone lesions (MOS);
2) The prevalence rates of bone diseases and their anatomical distribution *per* sex, laterality and age at death;
3) The correlation between lesions and the subsistence patterns of ancient people from Constância centred on fishing, boat construction and shipping goods through the Tagus and Zêzere rivers, and compared with other populations that may had a similar economy.

The research of MOS will be further supported by ethnographic and historical documents that will corroborate the skeletal evidences of occupation. Regarding the transport of goods (figure 20.3), the historical records reveal that in 1867, 60% of the males with 20/21 years old were "marítimos" and worked on boats; a life made by the rivers and from the rivers (Coelho, 1991).

The oral tradition, the material culture and the traditions, like the religious festivities of Nossa Senhora da Boa Viagem (figure 20.4), complete the sources, justifying an interdisciplinary approach.

Material and methods

In the years of 2002/2003, during the renewal of urban infrastructures in the historical centre of Constância, human skeletons were brought to the light. Regardless of the surprise, this was not a bizarre discovery, since the site was occupied by the ancient Church of São Julião and an old necropolis, dated from the XIV-XIX centuries. This finding lead to an archaeological intervention that made possible the recovery of 151 skeletons. For the present investigation a total of 64 skeletons, 42 males and 22 females, with know sex and age at death, were selected for analysis.

To study the MOS different methodological proposals will be adopted. The recording protocol of Buikstra e Ubelaker (1994) will be applied to the study of osteoarthrosis, concerning the analysis of degenerative bone lesions of the vertebral elements and peripheral articulations. In the vertebral body the presence of osteophytes or bony spurs formation will be evaluated. To the vertebral facets and main peripheral joints of the skeleton the occurrence of lipping, surface porosity and eburnation will be recorded.

The study of musculoskeletal markers (MSM) will be done by the method of Mariotti and co-authors (2004). This recording method evaluates the presence of robusticity, detectable by the presence of hypertrophy in particular points of attachment of muscles or ligaments; and/or by the formation of enthesopathies, in its dual occurrence: a proliferative or osteophytic form (OF), or an erosive or osteolytic form (OL). The analysis will be completed in a total of 36 muscles and ligaments

Fig. 20.3. The transport of goods in the boats (adapted from Coelho, 1998:15)

Fig. 20.4. Illustration of Nossa Senhora da Boa Viagem, the Maritimes protector (Coelho, 1998:18)

insertions of the upper/lower extremities and vertebral apparatus.

In the particular case of the stress fractures and postural markers, their presence or absence will be recorded. For all cases the bone severity, lateral distribution, sex and age at death will be taken in count.

PROSPECTS

The occupation is a crucial element in the social and cultural universe of people. However, the activity may have some effects on the health of individuals, in particular, if it demands the performance of repetitive movements. This is a true reality either for the present or for the past.

For ancient populations, the study of occupation and the subsequent impacts on health is possible through the analysis of markers of occupational stress (MOS), a distinctive condition of the skeleton that is produced in response to biomechanical stress. This issue rest under the goals of the described project. Starting from a skeletal sample, an interdisciplinary approach will be undertake combining the biological data, with historical, ethnographic and economic records that describe the main occupations presents during centuries in Constância.

With this research we will try to understand the impact of human activities in the skeleton, reconstructing a more reliable picture about the ancient inhabitants of Constância.

Acknowledgments

The author would like to thank to Grupo de Estudos em Evolução Humana (GEEvH) for their support and especially to Susana Carvalho for the previous draft of this paper.

References

CARDOSO, F. (2008) – *"A portrait of Gender in Two 19th / 20th Portuguese Populations: A paleopathological perspective"*. Academic Dissertation. Department of Archaeology. Durham: Durham University.

CARDOSO, H.; LOPES, L. (2002) – Skeletal markers of occupational stress in the toes: a case report from Alcabideche (Cascais, Portugal). *Antropologia Portuguesa*. 19: p. 19-28.

CAPASSO, L.; KENNEDY, K.; WILCZAK, C. (1999) – *Atlas of Occupational Markers on Human Remains*. Teramo: Edigrafital S.P.A.

COELHO, A. (1991) – *Festa de Nossa Senhora da Boa Viagem, a bênção dos barcos e o abraço dos Homens*. Constância: Câmara Municipal de Constância.

COELHO, A. (1998) – *Nos rios de Constância: a faina, a fé e a festa*. Constância: Câmara Municipal de Constância.

CUNHA, E.; UMBELINO, C. (1995) – What can bones tell about labour and occupation: the analysis of skeletal markers of occupational stress in the Identified Skeletal Collection of the Anthropological Museum of the University of Coimbra. *Antropologia Portuguesa*. 13: p. 49-68.

CUNHA, E. (2003) – Aproximación paleopatológica a algunas enfermedades reumáticas. In Isidro, A.; Malgosa, A., eds. – *Paleopatología: la enfermedad no escrita*. Barcelona: Masson SA, p. 209-220.

FERREIRA, L. (2006) – Análise *de vários indicadores esqueléticos para a reconstituição de padrões de actividade em populações humanas do passado – o caso de São Miguel de Odrinhas*. Trabalho de conclusão da Licenciatura em Biologia. Évora: Universidade de Évora.

HAWKEY, D.; MERBS, C. (1995) – Activity-induced musculoskeletal stress markers (MSM) and subsistence strategy among ancient Hudson Bay Eskimos. *International Journal of Osteoarchaeology*. 5: 4, p. 324-338.

IWAMOTO, J.; TAKEDA, T. (2003) – Stress fractures in athletes. *Journal of Orthopaedic Science*. 8: 3, p. 273-278.

KENNEDY, K. (1989) – Skeletal markers of occupational stress. In Iscan, M.; Kennedy, K., eds. – *Reconstruction of Life from the Skeleton*. New York: Alan R. Liss, p. 129-160.

KUMAR, S. (2001) – Theories of musculoskeletal injury causation. *Ergonomics*. 44: 1, p. 17- 47.

LARSEN, C. (1997) – *Bioarchaeology. Interpreting Behavior from the Human Skeleton*. Cambridge: Cambridge University Press.

LARSEN, C. (2000) – *Skeleton in our closet: revealing our past though Bioarchaeology*. Princeton: Princeton University Press.

MARIOTTI, V.; FACCHINI, F.; BELCASTRO, M. (2004) – Enthesopathies-proposal of a standardized scoring method and applications. *Collegium. Anthropologicum*. 28: 1, p. 145-159.

MAYS, S. (1998) – The *archaeology of human bones*. London: Routledge.

MERBS, C. (1989) – Trauma. In Iscan, M.; Kennedy, K., eds. – *Reconstruction of life from the skeleton*. New York: Alan Liss, p. 161-189.

NELSON, D.; CONCHA-BARRIENTOS, M.; DRISCOLL, T.; STEENLAND, K.; FINGERHUT, M.; PUNNETT, L.; PRÜSS-ÜSTÜN, A.; LEIGH, J.; CORVALAN, C. (2005) – The global burden of selected occupational disease and injury risks: methodology and summary. *American Journal of Industrial Medicine* (preprint version) [Em linha] [Consult. 16 Nov. 2006]. Disponível em WWW: _URL: http://www.who.int/quantifying_ehimpacts/methods/en/leigh.pdf_.

NIVA, M. (2006) – *Fatigue bone stress injuries of the lower extremities in Finnish conscripts*. Academic Dissertation. Department of Radiology. [em linha] Helsinki: University of Helsinki. [Consult. 16 Nov. 2006]. Disponível em WWW: _URL: http://ethesis.helsinki.fi/julkaisut/laa/kliin/vk/niva/fatigueb.pdf_.

QUEIROZ, M. (2000) – Evolução histórica das ideias sobre as doenças reumáticas. In Queiroz, M. eds. *Reumatologia*. Vol. I. Lisboa: Lidel, p. 3-7.

ROBERTS, C.; MANCHESTER, K. (2005) – *The archaeology of disease*. Gloucestershire: Sutton Publishing. 3rd Edition.

RODRIGUES CARVALHO, C. (2004) – *Marcadores de Estresse Ocupacional em Populações Sambaquieiras do Litoral Fluminense*. Tese de Pós-graduação em saúde pública. Escola Nacional de Saúde Pública: Fundação Oswaldo Cruz.

ROBB, J. (1994) – Skeletal signs of activity in the Italian metal ages: methodological and interpretative notes. *Human Evolution*. 9: 3, p. 215-229.

ROBB, J. (1998) – The interpretation of skeletal muscle site: a statistical approach. *International Journal of Osteoarchaeology*. 8: 5, p. 363-377.

STEEN, S.; LANE, R. (1998) – Evaluation of habitual activities among two Alaskan Eskimo populations based on musculoskeletal stress markers. *International Journal of Osteoarchaeology*. 8: 5, p. 341-353.

STIRLAND, A. (1998) – Musculoskeletal evidence for activity: problems of evaluation. *International Journal of Osteoarchaeology*. 8: 5, p. 354-362.

STEINBOCK, R. (1976) – *Paleopathological diagnosis and interpretation: bone diseases in ancient human population*. Springfield: Charles C. Thomas.

WILCZAK, C.; KENNEDY, K. (1998) – Mostly MOS: technical aspects of identification of skeletal markers of occupational stress. In Reichs, K., eds. – *Forensic osteology: advances in the identification of human remains*. Springfield: Charles C. Thomas Publisher, p. 461-490.

ANÁLISE ANTROPOLÓGICA DOS VESTÍGIOS OSTEOLÓGICOS DE UM POSSÍVEL CASO DE GRAVIDEZ EM FIM DE TEMPO (ADRO DA IGREJA ANTIGA DO OLIVAL-OURÉM)

C. CRUZ & R. MARQUES

Grupo de Estudos em Evolução Humana, Departamento e Antropologia da Universidade de Coimbra

Abstract: *The Olival's church (Ourém) had, through out the time, an important amount of relevance in the regional context. Despite that, during the second half of the last century it was neglected to a point near the complete destruction. In 2005-2006 the municipality and the Parrish jointed forces to reconstruct and recover both the church and the religious art pieces. For that archaeological, as well as anthropological technical expertise was required. The archaeological survey stared at the adro (the open space that surrounds the church) where several burials were found. One of the burials showed some particular and atypical features: in the pelvic region of a feminine skeleton, were found bones of an infant revealing, some of them, an anatomical connection, leading us to hypothesise about the relation of the two bone assemblage as being of a mother and child. Given the scarcity of such evidence through the prehistoric and historic periods it is important to discuss such cases.*
Key words: *Pregnancy; Paleopathology; Birth; Cause of death; Infant*

Résumé: *L'église de L'Olival a Ourem a toujours eu au long des années une grande importance dans la région. Cependant, aux années 60, les activités de culte a son intérieur ont cessée. Cette circonstance a précipitée sa détérioration.*
En 2005-2006 la mairie et la commission administratif de l'église ont débutée des travaux d'amélioration vis-à-vis sa structure architecturel, les œuvres d'art a son intérieur, ainsi qu'aux fouilles archéologiques et anthropologiques. Les fouilles archéologiques ont commencé aux extérieurs de l'église. Plusieurs vestiges des funérailles y ont été découverts, un de ceux appartenant a l'ensemble d'un individu du sexe féminin et un enfant. Les motivations pour ce fait rare restent mystérieuses, mais le décès d'une femme en la phase finale de sa grossesse pourrait peut être l'expliquer.
Le caractère très rare de ce genre de trouvaille doit stimuler notre recherche vis-à-vis des causes d'une telle mort, comme un moyen de jeter de la lumière sur ce genre de décès, apparemment très commun entre les populations anciennes.
Mots-clés: *Grossesse; Paléopathologie; Naissance; Cause de Décès; Infant*

Resumo: *A Igreja do Olival – Ourém, teve ao longo do tempo uma importância considerável na região. Contudo, na década de 60 do século passado foi abandonada ao culto tendo desde essa altura vindo a degradar-se rapidamente, causando a quase completa destruição do seu interior. Em 2005-2006 a Câmara Municipal e a Comissão Fabriqueira da Igreja deram início às obras de melhoramento no que diz respeito às peças de arte sacra, a arquitectura e aos vestígios arqueológicos e antropológicos. A intervenção arqueológica teve início com uma sondagem de prospecção aberta no Adro da Igreja. Foram encontrados vários enterramentos, sendo que um deles pertence a um indivíduo do sexo feminino inumada conjuntamente com um infante não sendo clara a motivação destes enterramentos, pode tratar-se da inumação de uma mulher em termo de gravidez. Dada a raridade deste tipo de achados em arqueologia é importante avaliar e discutir vários aspectos relativos a este enterramento já que este caso pode ajudar a compreender vários aspectos relacionados com esta causa de morte comum entre as populações do passado.*
Palavras-chave: *Gravidez; Paleopatologia; Nascimento; Causa de Morte; Infante*

INTRODUÇÃO

A escassez de dados e evidências arqueológicas relacionadas com mulheres em final de tempo de gravidez, faz com que os novos dados que possam surgir, se tornem fundamentais para o conhecimento mais profundo desta etapa muito particular da vida das mulheres. Por este motivo julgamos importante a inclusão deste caso, que apesar de não ser pré ou proto-histórico, poderá servir de ferramenta interpretativa de possíveis casos de inumação de mulheres grávidas em épocas mais remotas.

O nascimento entre os humanos segue um padrão único e distinto entre os mamíferos em geral e entre os primatas em particular. As principais diferenças assentam no facto de que este é um processo biológico bastante complexo e laborioso que inclui a orientação que o feto assume ao passar pelo canal pélvico e o modo como ele emerge através do mesmo. Estas diferenças são complementadas pela carga cultural do papel social desempenhado pela mãe e por outros indivíduos que prestam a sua ajuda no momento do nascimento. Neste contexto, o nascimento em humanos tem importantes implicações para a morfologia do osso coxal e por inerência para a estrutura da cintura pélvica, acentuando assim o dimorfismo sexual ao nível desta zona anatómica. Ao mesmo tempo, outras características emergem tal como o comportamento de cooperação social que neste complicado momento do ciclo de vida reprodutivo humano permite a terceiros facilitar a chegada de novos membros da sociedade, assim como proporcionar à mulher os cuidados necessários à convalescença pós-parto.[1]

O nascimento humano moderno evoluiu, tal como outras características distintivas da nossa espécie, seguindo um padrão de tipo "mosaico", significando esta expressão que algumas das características e particularidades surgiram

[1] Rosenberg, 1992.

Fig. 21.1. Destaca a ponteado da área da sondagem 1

relativamente cedo na nossa história evolutiva e outras bastante mais tarde. Assim, de acordo com a interpretação que nos chega dos testemunhos fósseis, o padrão moderno de aproximação da cabeça do feto ao canal de parto estaria já presente entre os australopitecíneos, há cerca de 3 milhões de anos atrás, sendo esta ilação retirada a partir da análise morfológica métrica dos restos fósseis da cintura pélvica atribuíveis a indivíduos deste género. Por outro lado, outros aspectos do nascimento humano moderno incluem a rotação da cabeça, tronco e resto do corpo do bebé enquanto atravessa o canal de parto, assim como a distinta posição de emergência da cabeça na fase final do processo. Estas particularidades surgiram provavelmente já mais recente mente e em resposta à crescente encefalização das espécies pertencentes ao género *Homo* nossas antecessoras, na exacta medida em que este facto se fez acompanhar de pressões evolutivas que conduziram tanto à alteração da forma da pélvis como também a diferenciações intrínsecas do processo e dos momentos de nascimento.[2]

CONTEXTO ANTROPOLÓGICO / CONTEXTO DE ESCAVAÇÃO

Os dados relativos à intervenção arqueológica serão descritos sumariamente, uma vez que no relatório técnico-científico (em arquivo no Instituto Português de Arqueologia) se referir mais pormenorizadamente os detalhes de carácter arqueológico da escavação.[3]

A intervenção arqueológica/antropológica ocorreu na zona do adro da Igreja Antiga do Olival, onde foi aberta uma sondagem com cerca de 152 cm – 190 cm de largura por 258 cm – 320 cm de comprimento (Figura 21.1). Foram encontradas algumas estruturas em pedra sob as quais existia uma camada de preparação de terra e uma camada de ossos dispersos sem qualquer conexão entre eles. À profundidade média de 80cm (em relação à soleira da Igreja) surgiram os primeiros enterramentos, entre os quais se salienta o caso de uma mulher que à morte poderia estar no final de tempo de gravidez.

Para além dos enterramentos, foram também encontrados alguns vestígios arqueológicos, sobretudo moedas (sendo numa delas visível uma referência ao reinado de D. Maria II) e contas de rosários que permitem datar estes enterramentos com tendo ocorrido durante a época moderna.[4]

Antes de mais, importa referir que todos os elementos que foram e vão ser referidos, foram recolhidos em contexto de escavação. Desta forma, o processo de recolha de

[2] Rosenberg, 1992.

[3] Pereira, 2006.

[4] Pereira, 2006.

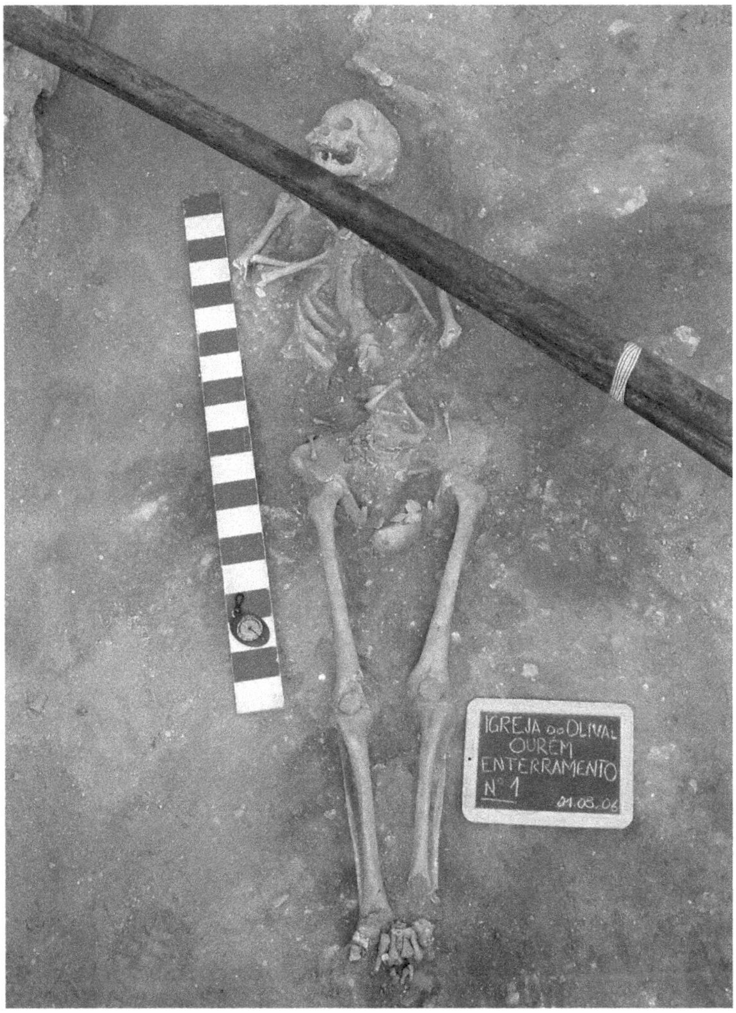

Fig. 21.2. Enterramento 1 *in situ* após limpeza e exposição

Tab. 21.1. Dados métricos do Enterramento 1

Tipo de osso	Lado esquerdo	Lado direito	Observações
Comprimento Clavícula	120 mm*	128 mm	* osso fragmentado
Comprimento Úmero	245 mm*	230 mm*	* osso fragmentado
Comprimento Rádio	200 mm	190 mm*	* osso fragmentado
Comprimento Fémur	395 mm	395 mm	
Diâmetro transversal da cabeça	35 mm	35 mm	Permite a diagnose sexual segundo Wasterlaine, 2000

dados, está inevitavelmente condicionado por constrangimentos de tempo, das condições climatéricas e recursos técnicos, o que pode conduzir a avaliações que, apesar de rigorosas, podem ser sub ou sobre valorizadas. Na medida em que não foi requisitada uma análise antropológica laboratorial, os dados recolhidos em campo foram utiliza-dos para a análise que a seguir se apresenta.

Sobre o Enterramento 1 (designação de campo atribuída à inumação do caso em análise) os dados recolhidos em campo permitem concluir, no que à Antropologia funerária diz respeito, que este indivíduo teria sido inumado em decúbito dorsal de acordo com a norma canónica (crânio – Oeste / pés – Este). O crânio encontrava-se ligeiramente descaído sobre o lado direito, estando os membros superiores flectidos sobre a zona torácica com as mãos sobrepostas sobre o esterno. Os membros superiores encontravam-se esticados e paralelos entre si. O comprimento máximo deste esqueleto era de cerca de 155 cm, sendo a largura máxima dada pela zona pélvica de 36 cm[5] (Figura 21.2).

[5] Cruz, 2006.

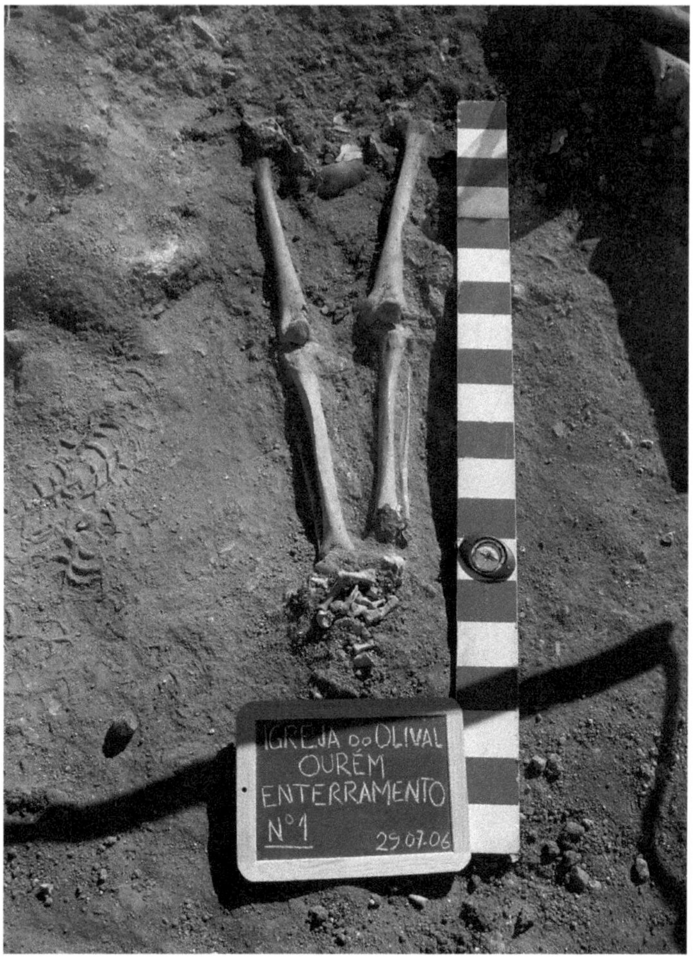

Fig. 21.3. Enterramento 1 no momento em que foram expostos os primeiros ossos de não-adulto na região pélvica

A estimativa da idade à morte foi feita tendo por base a metodologia proposta por Brothwell[6] que determina uma idade média para a fusão das epífises e permite apontar uma idade superior a 25-30 anos, já que todos os centros de ossificação apresentavam este processo concluído.

A diagnose sexual foi realizada, tendo em conta o método proposto por Ferembach[7] que se baseia nas características morfológicas de determinadas regiões do esqueleto: crânio e osso coxal. Esta análise, apesar do estado de preservação do material, permitiu no geral defender que se trataria de um indivíduo do sexo feminino. A análise métrica, desenvolvida por Wasterlain[8] foi realizada apenas na cabeça do fémur, sendo o valor (35 mm) inferior ao do ponto de cisão, indicando que se trataria de um indivíduo do sexo feminino.[9]

De acordo com uma avaliação macroscópica foi apenas identificada uma possível fusão de duas vértebras torácicas (T9 e T10) bem como indícios de osteofitose ao nível da L1. Para além destas não foram detectadas quaisquer outras evidências da existência de patologias que possam deixar marcas no esqueleto. Contudo ao nível da patologia oral, vale a pena salientar que se verificou a reabsorção alveolar dos segundos pré-molares e dos três molares inferiores de todos os quadrantes.[10]

No momento da limpeza e exposição do Enterramento 1 (Figura 21.3 e Figura 21.4), foram encontradas algumas peças osteológicas pertencentes a um indivíduo sub-adulto que se estendiam desde um pouco abaixo da sínfise púbica até à região lombar. Uma análise cuidada em campo, permitiu perceber que alguns destes pequenos ossos se encontravam um pouco dispersos, nomeadamente os ossos longos do infante, facto que poderá estar relacionado com a acção de agentes tafonómicos,[11] mas entre as costelas e os processos vertebrais existia uma associação anatómica e encontravam-se assentes sobre o sacro do indivíduo adulto. Por este motivo foi-lhe atribuída a designação de Enterramento 1B. Na zona pélvica do indivíduo do Enterramento 1 encontravam-se alguns ossos da abóbada

[6] Brothwell, 1981.
[7] Ferrembach, 1979.
[8] Wasterlain, 2000.
[9] Cruz, 2006.
[10] Cruz, 2006.
[11] Cruz, 2006.

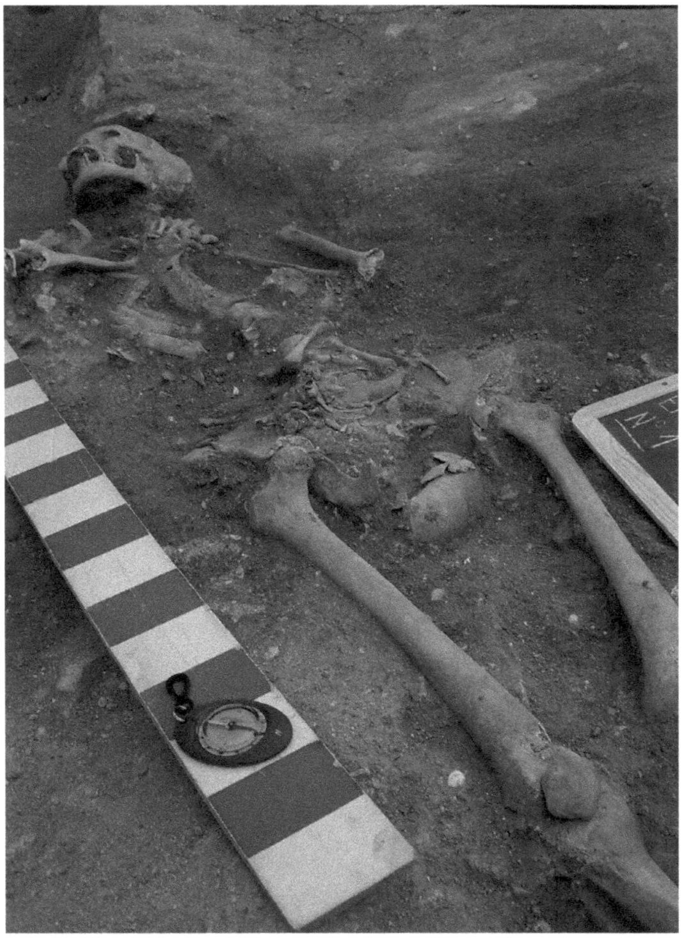

Fig. 21.4. Enterramento 1*in situ* após exposição completa. É visível na zona pélvica do indivíduo adulto ossos do infante

Tab. 21.2. Dados métricos do Enterramento 1B

Tipo de osso	Lado esquerdo	Lado direito	Observações
Comp. Úmero	62* mm	62* mm	* osso fragmentado
Comp. Cúbito	59 mm	*	* presente apenas o oleocraneo
Comp. Fémur	73,5 mm	73,5 mm	
Comprimento Tíbia	66,5 mm	66,5 mm	

craniana do indivíduo sub-adulto, sendo que os ossos da face se encontravam voltados e no interior do substrato. O comprimento dos ossos longos, deste indivíduo não-adulto permitiu a estimativa da idade à morte entre 36 a 40 semanas lunares,[12] que coincidem com o tempo normal de uma gestação em fim de tempo.

A análise macroscópica dos ossos permitiu observar sobreposição de osso novo, podendo esta ser uma característica associada ao intenso desenvolvimento osteológico que os neo-natos experimentam, ou uma reacção do osso a um quadro infeccioso[13] (Figura 21.5 e Figura 21.6). No caso deste último pressuposto ser verdadeiro, pode este eventual quadro patológico ser um possível causador do fracasso deste nascimento.

A existência de ossos de um indivíduo não-adulto na região pélvica, associada à posição do crânio e da permanência de uma relação anatómica de algumas peças osteológicas, à diagnose sexual do indivíduo do Enterramento 1 e à estimativa da idade à morte do não-adulto, permite levantar a hipótese, de se tratar de um enterramento de mãe e filho, sendo a uma primeira análise plausível a interpretação do não nascimento do indivíduo não adulto atrás descrito[14] (Figura 21.7).

[12] Fazékas e Kósa, 1978 *in* Scheuer e Black, 2000.

[13] Cruz, 2006.

[14] Cruz, 2006.

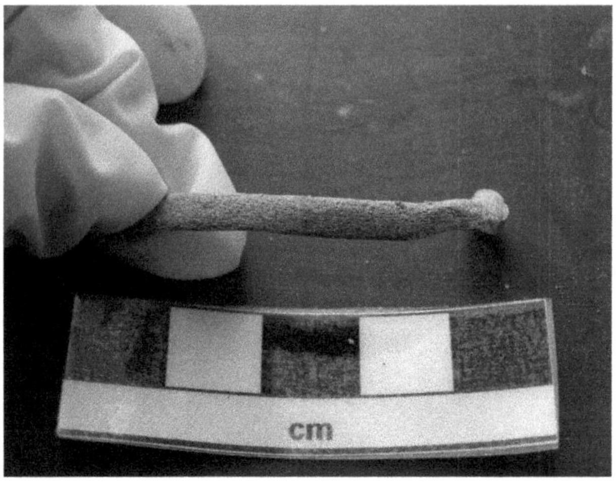

Fig. 21.5. Costela do indivíduo sub adulto (Enterram ento 1B) em que é visível a porosidade associada à intensa actividade reactiva do osso

Fig. 21.6. *Pars basilaris* com porosidade associada à actividade reactiva do osso

Fig. 21.7. Detalhe da zona pélvica do Enterramento 1 onde são visíveis os ossos do indivíduo sub-adulto

ALGUMAS CONSIDERAÇÕES SOBRE ESTE ENTERRAMENTO

Sobre a relação biológica e/ou social entre estes dois indivíduos e o que de facto poderá ter acontecido relativamente ao não-nascimento ou nascimento desta criança, é apenas possível levantar hipóteses baseadas nos dados recolhidos em campo.

A posição do crânio parece coincidir com aquela que se encontra no momento do parto, isto é, os pés orientados em direcção ao crânio da mãe e a face em posição posterior em relação ao corpo da mãe. A idade estimada para o feto coloca-o perto do momento do nascimento, ou até pouco depois (36 a 40 semanas lunares).[15] Ao nível da gestação e em casos de normalidade de desenvolvimento intra-uterino, o termo da gravidez está associado a uma mudança de posição do feto que coincide com a posição encontrada no local da escavação e que já foi anteriormente descrito.

No caso da criança ter nascido, a posição escolhida para o enterramento pode parecer um pouco estranha, a menos que com ela se procurasse replicar os últimos momentos

[15] Scheuer e Black, 2000.

de ambos os indivíduos. Esta interpretação atribui um papel relevante ao nível social, cultural e afectivo que é espectável numa situação repleta emoções e de expectativa a que está associado o nascimento de uma criança.

COMPLICAÇÕES ASSOCIADAS À GRAVIDEZ

O não nascimento da criança, o que poderá ser explicado por um conjunto de patologias ou acontecimentos que podem comprometer o momento do nascimento. A par da puberdade, a gravidez é porventura o evento mais significativo na vida da mulher nas suas vertentes biológica e cultural, na exacta medida em que, se por um lado se assiste a várias alterações em termos fisiológicos do corpo humano, por outro, é este o momento em que é preparado o papel social de mãe. Uma vez que para a presente análise nos interessa sobretudo a vertente biológica do evento, cabe dizer que ao nível da cintura pélvica, mais especificamente no que diz respeito aos ossos coxais, são várias e profundas as alterações na morfologia óssea que ocorrem; de forma tal que na década de 60 se tentou contabilizar o número de partos que uma mulher teria feito através das modificações, que destes ossos. As alterações seriam sobretudo perceptíveis ao nível da sínfise púbica, em que as alterações são de carácter muscular, ligamentar, articular e não articular, uma vez que o alargamento do canal de parto provoca a distensão anormal de músculos e o esforço extraordinário de ligamentos, eventos auxiliados pela produção de um ambiente fisiológico específico e pela libertação de substâncias químicas que facilitam o processo. No entanto, todo este stress deixa marcas sobretudo sob a forma de exostoses e depressões na superfície óssea.[16] Outros autores sistematizam mais estas alterações: subsistência de fossas na região púbica dorsal, sulcos pré-auriculares e sulcos localizados em regiões de inserção de ligamentos interósseos.[17]

São também conhecidas algumas descrições que atestam a relativa fragilidade em termos imunitários que a gravidez acarreta. Com efeito, existem evidências para um aumento de inflamação na gengivas de mulheres grávidas, sendo que esta condição, a par de uma fraca higiene oral, presença de cáries, stress nutricional ou psicológico, são factores de peso na presença de doenças como por exemplo a periodontite aguda e consequente perda de dentes *ante-mortem*.[18]

Por todas estas alterações e riscos, a gravidez acarreta grandes probabilidades de aumentar a morbilidade e a mortalidade das mulheres nesta condição, sendo que nos dias que correm, morrem por ano pelo menos meio milhão de mulheres grávidas em países em desenvolvimento.[19] É possível que o cenário entre os séculos XV e XIX, período em que muito provavelmente viveram a mulher e o não adulto em questão, fosse ainda mais grave. Esta afirmação é suportada pelo facto de que na Europa entre o Séc. XVII e XX as mortes por obstrução do canal de parto eram elevadas devido à ampla dispersão do raquitismo. Estes problemas podem ter sido agravados tendo em mente que as necessidades alimentares dos homens eram satisfeitas primeiramente do que as das mulheres, sobretudo em momentos de escassez.[20]

Tendo em mente todos estes riscos e contingências, podemos apontar como possíveis algumas complicações que podem surgir durante o parto e que podem provocar a morte tanto do adulto como do não adulto.[21]

a) obstrução do canal de parto, que pode resultar das reduzidas dimensões deste ou pelo facto de que o bebé se apresenta em posição anormal à altura do nascimento;

b) infecção contraída pela mãe, sendo esta condição resultante do facto de que a mãe tem as suas defesas imunitárias reduzidas por forma a que o seu corpo não rejeite o feto em desenvolvimento. Simples infecções urinárias podem apresentar graves riscos para mãe e filho;

c) hemorragias durante o nascimento, que pode ser provocada por uma variedade de factores, mas cuja causa principal que afecta 5% dos casos é uma deficiente implantação da placenta no útero. Associada a esta, outra complicação pode surgir, esta menos bem compreendida: a ecplampsia. Esta forma de toxémia é caracterizada por convulsões e/ou coma.

COMENTÁRIOS FINAIS

Apesar de não ser possível afirmar que o caso em análise se constitui numa morte de mãe e filho durante o trabalho de parto, podemos pelo menos assumir que as complicações acima descritas se assumiram como muito mais gravosas nas sociedades do passado em que não estariam ainda desenvolvidas as modernas práticas terapêuticas e de intervenção cirúrgica.

Como foi já referido, os ossos do infante apresentavam sinais de uma intensa actividade que se reflecte com a deposição de osso novo. Este evento, entre indivíduos em desenvolvimento nem sempre é claramente percebido, isto porque a actividade pode tanto estar associada ao crescimento, quanto a uma resposta a uma condição patológica de origem infecciosa, que associado ao processo de crescimento, resulta o mais das vezes numa resposta exuberante. Importa contudo assinalar que, na eventualidade de se tratar do enterramento de uma mulher ainda grávida, a causa de morte de mãe e feto poderia estar associada a uma possível enfermidade do feto que ao se agravar colocou em causa a saúde da mãe.

[16] Ubelaker, 1974.
[17] Kelley, 1979.
[18] Larsen, 1997.
[19] Ortner, 2003.
[20] Ortner, 2003.
[21] Ortner, 2003.

São apenas hipóteses as que surgem quando um caso como este é analisado, contudo existem estratégias que se podem desenvolver para corroborar ou descartar as hipóteses descartadas. Uma vez que o enterramento é de uma época recente, a consusta dos resgistos paroquiais pode fazer referência a um caso que apresente elementos que vão ao encontro do que foi anteriormente apresentado. Por outro lado, uma análise ao material genético de ambos indivíduos, permitiria pelo menos inferir sobre o parentesco entre os dois. Seja qual for a realidade por de trás deste caso, ele mostra-se importante para compreender tanto os processos biológicos associados ao parto quanto os processos culturais, podendo a partir deste ponto extrapolar para casos mais remotos.

Bilbiografia

BAPTISTA, A.R. (1982) – *Conta dos oficiais da Câmara de Ourém sobre o terramoto de 1755.* Estudos e documentação, p. 95-100.

BROTHWELL, D.R. (1981) – *Digging up bones: the excavation, treatment and study of human skeletal remains.* London: British Museum.

CRISTINO, L.C. (2003) – D. Afonso 4º Conde de Ourém e a Fundação da Colegiada. In *D. Afonso 4º Conde de Ourém e a sua época.* Ourém: Câmara Municipal de Ourém.

CRUZ, C.B. (2006) – *Relatório Antropológico de Campo: Igreja Antiga do Olival (Ourém).* Ourém: Câmara Municipal de Ourém.

FEREMBACH, D., I. SCHWIDETZKY & M. STLOUKAL. (1980) – Recommendations for age and sex diagnosis of skeletons. *Journal of Human Evolution.* 9: p. 517-549.

KELLEY, M.A. (1979) – Parturition and pelvic changes. *American Journal of Physical Anthropology.* 51: p. 541-545.

LARSEN, C.S. (1997) – *Bioarchaeology: interpreting behavior from the human skeleton.* Cambridge: Cambridge University Press.

OLIVEIRA, J.F. (2003) – *D. Afonso e as origens de Ourém.* Paper presented to the D. Afonso: 4º Conde de Ourém, e a sua época, Ourém, 2003.

ORTNER, D.J. (2003) – Identification of Pathological Conditions in Human Skeletal Remains. London, U.K: Academic Press.

PEREIRA, J. (2006) – *Relatório Arqueológico preliminar: Igreja Antiga do Olival.* Ourém: Câmara Municipal de Ourém.

ROSENBERG, K.R. (1992) – The evolution of modern human childbirth. *American Journal of Physical Anthropology.* 35: p. 89-124.

SCHEUER, L. & S. BLACK. (2000) – *Developmental Juvenile Osteology.* USA: American Press.

UBELAKER, D.H. (1989) – *Human skeletal remains: excavation, analysis interpretation.* Washington: Taraxacum.

WASTERLAIN, R.S.C.N. (2000) – *Morphé: análise das proporções entre os membros, dimorfismo sexual e estatura de uma amostra da colecção de esqueletos identificados do Museu Antropológico da Universidade de Coimbra.* Dissertação de Mestrado em Evolução Humana. Coimbra: Universidade de Coimbra.

WORKING MEMORY AND MODERN HUMAN MIND

Manuel MARTÍN-LOECHES
Center for Human Evolution and Behavior, UCM-ISCIII. Sinesio Delgado, 4, Pabellón 14, 28029 Madrid, Spain.; Tel.: + 34 91 387 75 43. Fax: + 34 91 387 75 48. Email: mmartinloeches@isciii.es; URL: www.ucm.es/info/neurosci

***Abstract**: From the Cognitive Neurosciences and the Psychology, it is being recently proposed that a single additive genetic mutation might have yielded enhanced working memory (EWM). As a consequence, modern mind emerged from an already anatomically modern brain by virtue of a relatively simple mechanism. In the present work, the proposal is explained, defining what working memory is and how a mere enhancement of this process could have yielded modern human mind.*
***Key words**: Brain, Evolution, Working Memory, Mind, Language*

INTRODUCTION

What we are going to bring out here is what could have been the mechanism by virtue of which modern mind, our mind, emerged. A mind that made us able to win the race when faced to Neanderthals in Europe, most probably because this mind was more capable of exploiting the natural resources and successfully compete to attain them. We will also comment how could this fact have easily succeeded in a brain that was nonetheless anatomically modern several thousands of years before it yielded modern mind. That is, our starting point is that the brain of the *Homo sapiens* could have reached its anatomical properties much time (prior to 150 Ka) before a plainly modern mind came out from such a brain (after 100 Ka, may be much later). This is at least a possibility, even if an alternative possible scenario would be the emergence of both anatomically modern brain and modern mind, concurrently, prior to 150 Ka.

The mechanism has been proposed to be a very simple one, though with dramatic consequences. This mechanism has been called Enhanced Working Memory (EWM), and two main authors proposing and supporting this hypothesis are Frederick Coolidge and Thomas Wynn (2005, Wynn and Coolidge, 2004). Basically, the hypothesis proposes that our working memory capacity was enhanced and that, as a consequence, all the modern human faculties would have easily arisen including language, thought, and so on. Enhancing working memory capacity, in turn, does not need dramatic changes in the brain, but subtle ones. However, to plainly understand this proposal, we have first to introduce the reader to the concept of working memory, as much as to how could and enhancement of this process have produced modern mind.

WHAT IS WORKING MEMORY?

First, reflect on the following text by the comedian Sue Murphy (cited in Calvin 2004): "*Did your ever walk into a room and forget why you walked in? That's how dogs spend their lives*". This is an example of poor working memory.

An easy way to define working memory could be to call it "what we have in mind" for a while. What we have in mind may come from either our perceptual organs (what we see, what we hear, etc.) or from our long-term memory, from our past experiences, and most times it is a mixture of both sources. Sometimes working memory has been identified with "consciousness", though this is somewhat controversial.

Consciousness and working memory are not necessarily the same. Working memory may comprise both, elements of information of which we are conscious but also elements of information that are not necessarily activated to reach the consciousness level. That is, working memory would also include information of which we are not conscious, information that is sub-activated, activated to some extent but not so much as the information of which we are conscious. This sub-activated information can easily be brought to consciousness, as when we hear a word (e.g., "boat") and we can effortlessly be conscious of any of the features related to that word (e.g., it serves to navigate, it can have several shapes and sizes, it has an anchor, and so on), though only some of those features will be activated to the consciousness level. Accordingly, working memory is information that we have in mind, even if part of this information is not totally conscious but may become conscious easily.

An important feature of working memory is included in its name: it is a memory to work. It is a working space. Information is not only activated or sub-activated, but we actively work with that information. Indeed, the better way to understand this aspect is probably using as example one of the most standard ways to measure working memory capacity. Consider the following numbers: "8-5-4-1-7". Read them and, closing your eyes, try to repeat them in reversed order. This is working memory: several numbers entered some type of "buffer", our working memory, where we have moved them (worked with them) to a different order. If we increase the number of elements (in this case, the number of numbers) we will probably get in trouble, being this good example of the fact that our working memory is of limited capacity. Even if working memory has been enhanced in our species, it is still limited. Usually the mean number of

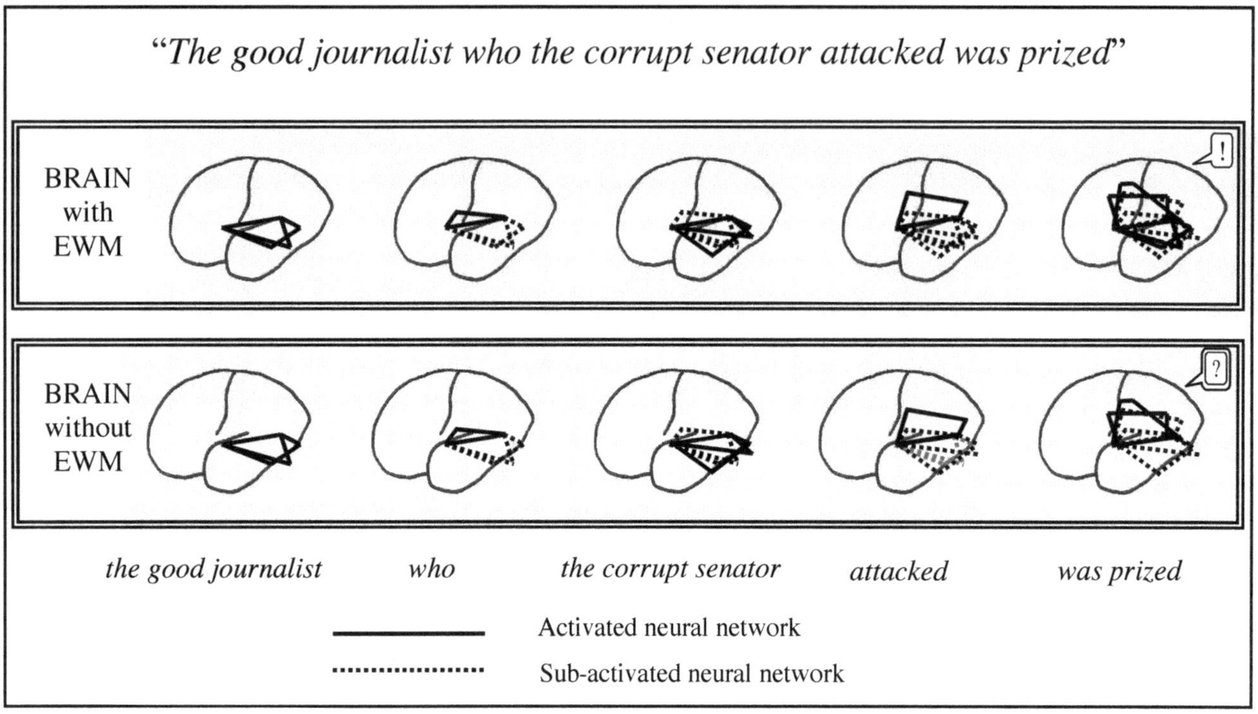

Fig. 22.1. Brains processing a relatively complex sentence. The brain in the upper panel has enhanced working memory (EWM) and therefore is able to keep activated and sub-activated simultaneously a larger number of neural networks representing the words of the sentence. This brain understands the sentence. The brain in the bottom panel, without EWM, does not understand the sentence since the possible number of activated and sub-activated networks is reduced

elements that can be reversed without failures is 5 (this can vary). Consider that, although somewhat speculative, before the enhancement of working memory this number could be lowered to 2 or 3, as would have been the case for anatomically modern humans before EWM, or even for our closest relatives the Neanderthals.

HOW COULD EWM EXPLAIN MODERN MIND? THE CASE OF LANGUAGE

With an enhancement in working memory capacity, human language can significantly be improved, achieving the unique features that characterize modern human language. How can EWM improve human language? A main central process in human language is to understand sentences. A sentence is composed by several words and the human brain's goal is to determine which are the relationships between these words, these relationships based upon a structure with some degree of hierarchy. In the sentence "*The little cat is beside the big dog*", the adjective "little" refers to cat and "big" refers to dog, both animals have some type of relationship (one is beside the other), but we are mainly speaking about one of them (the little cat, the subject of the sentence). This sentence could nevertheless be probably understood with a working memory capacity of a Neanderthal. But now consider the following sentence: "*The good journalist who the corrupt senator attacked was prized*". There is a relatively long distance between the "was prized" and "the good journalist", and some "disturbing" elements in between: another participant (the corrupt senator) and another verb (attacked), regardless of which we have no trouble in assigning "the good journalist" as the subject of the sentence and "was prized" as the main verb in the sentence, corresponding to our subject. This sentence could not be understood by a Neanderthal. The sentence is longer and the relationships between the elements are not linear. To understand a sentence as the latter one, as much as notably longer and more complex examples that could be brought here, we need the working memory capacity of modern humans. With EWM we can keep activated or sub-activated several elements in working memory until they are attached or related to their corresponding counterparts (in the example, the subject with the main verb).

Figure 22.1 represents, albeit extremely schematically, the differences between a brain with EWM and a brain without EWM in the processing of the later exemplified sentence. To suitably understand that sentence requires a certain number of neural networks to be kept activated and sub-activated. In the brain with EWM the number of neural networks activated and sub-activated can be larger and the neural network for the subject of the sentence can be re-activated in order to relate it to the just-appeared main verb of the sentence. A brain without EWM would be unable to reactivate the neural network of the subject, and even it is possible that this network could be completely deactivated at that time. This brain would not understand the sentence.

Then, if we have a better language, a more complex language that can define more complex relationships of the real world, we have a tremendously powerful tool to comprehend the complexities of the real world. In other words, we would have a better though. And here we would have a notably adaptive advantage to afford the natural environment. This is the proposal of several authors, as Jackendoff (1999, 2002).

But there are other proposals on the relationships between modern human language and modern capacity of reasoning, or modern though. As an instance, Aboitiz *et al.* (2006) rather propose that a general increase in working memory capacity would be a general purpose mechanism that would explain both a better reasoning capacity as much as modern language. Whatever the case, EWM appears a parsimonious hypothesis to explain modern human mind.

HOW COULD EWM HAVE OCCURRED?

According to neural models of working memory, a simple way to define an enhancement in working memory would be that the brain is able to keep activated and/or sub-activated a larger number of neural networks and for a longer time. To achieve this goal we have two possibilities.

Several authors propose that the enhancement in our cognitive abilities may be due to cultural factors (e.g., Calvin 2004). In this sense, each generation improves the capacities of the previous generation because its members have been exposed to knowledge at very early periods in their life. Early expositions to knowledge would help to keep alive a larger number of neurons, then improving and increasing neural connections. When the generation grows, more complex neural networks have been accomplished, which might in turn imply that a more complex knowledge has been achieved. This complex knowledge is passed early to the next generation, and the cycle continues. Such type of proposals implies a relatively gradual evolution of EWM. Whereas this might be acceptable, the problem would be to resolve why this did not occur to other species within the human lineage.

A second possibility is that a genetic mutation occurred. This view is shared by a certain number of authors (e.g., Coolidge & Wynn, 2005; Hauser *et al.*, 2002). Indeed, to attain an EWM a single additive genetic mutation appears sufficient. The only we need is an increase in the amount of neural tissue that can be activated simultaneously and for a longer time. Hence, we do not need to appeal to an absolute brain size increase nor even to any type of brain reorganization (in the sense of modifying the shape of the neural circuits) in order to attain a modern mind. We even do not need to increase the number of neurons. In this way, an already anatomically modern brain could become functionally modern by simply increasing the number of neurons that can be activated in parallel.

This enhanced number of simultaneously active neurons could be achieved by at least two non-mutually excluding means (Martín-Loeches, 2006). One, which nevertheless implies some degree of anatomical change and reorganization, is that the overall number of existing neurons for a given function is augmented. The other is that, keeping constant the number of neurons, some metabolic or other functional cell features permit more neurons to be activated simultaneously.

Several genetic loci appear, again, as possible, non-mutually excluding, candidates for these changes. One of best of these candidates and that suitably accounting accounts for both mechanisms is FOXP2. It has been shown that anomalies of this gene may cause both structural (number of existing neurons) and functional (number of activated neurons) abnormalities into certain brain regions related to language (Vargha-Khadem *et al.*, 1998). But other candidates also exist.

References

ABOITIZ, F. [*et al.*] (2006) – Cortical memory mechanisms and language origins. *Brain and Language, 98*, 40-56.

CALVIN, W.H. (2004) – *A brief history of the mind.* Cambridge University Press, Cambridge.

COOLIDGE, F.L., WYNN, T. (2005) – Working memory, its executive functions, and the emergence of modern thinking. *Cambridge Archeological Journal 15*, 5-27.

JACKENDOFF, R. (1999) – Possible stages in the evolution of the language capacity. *Trends in Cognitive Sciences 3*, 272-279.

JACKENDOFF, R. (2002) – *Foundations of Language: Brain. Meaning, Grammar, Evolution.* Oxford University Press, Oxford.

HAUSER, M.D. [*et al.*] (2002) – The faculty of language: What is it, who has it and how did it evolve? *Science 298*, 1569-1579.

WYNN, T., COOLIDGE, F.L. (2004) – The expert Neandertal mind. *Journal of Human Evolution 46*, 467-487.

VARGHA-KHADEM, F. [*et al.*] (1998) – Neural basis of an inherited speech and language disorder. *Proceedings of National Academy of Sciences (USA) 95*, 12695-12700.

MARTÍN-LOECHES, M. (2006) – On the uniqueness of humankind: is language working memory the final piece that made us human? *Journal of Human Evolution 50*, 226-229.

www.ingramcontent.com/pod-product-compliance
Lightning Source LLC
LaVergne TN
LVHW071213210926
848482LV00015B/797